T0155632

NEUROMETHODS

For further volumes:
http://www.springer.com/series/7657

Neurotransmitter Transporters

Investigative Methods

Edited by

Heinz Bönisch

Biomedical Center, Institute of Pharmacology and Toxicology, University of Bonn, Bonn, Germany

Harald H. Sitte

Center of Physiology and Pharmacology, Institute of Pharmacology, Medical University of Vienna, Vienna, Austria

Editors
Heinz Bönisch
Biomedical Center
Institute of Pharmacology and Toxicology
University of Bonn
Bonn, Germany

Harald H. Sitte
Center of Physiology and Pharmacology
Institute of Pharmacology
Medical University of Vienna
Vienna, Austria

ISSN 0893-2336 ISSN 1940-6045 (electronic)
Neuromethods
ISBN 978-1-4939-8132-8 ISBN 978-1-4939-3765-3 (eBook)
DOI 10.1007/978-1-4939-3765-3

Printed on acid-free paper

This Humana Press imprint is published by Springer Nature
The registered company is Springer Science+Business Media LLC New York

Series Preface

Experimental life sciences have two basic foundations: concepts and tools. The *Neuromethods* series focuses on the tools and techniques unique to the investigation of the nervous system and excitable cells. It will not, however, shortchange the concept side of things as care has been taken to integrate these tools within the context of the concepts and questions under investigation. In this way, the series is unique in that it not only collects protocols but also includes theoretical background information and critiques which led to the methods and their development. Thus, it gives the reader a better understanding of the origin of the techniques and their potential future development. The *Neuromethods* publishing program strikes a balance between recent and exciting developments like those concerning new animal models of disease, imaging, in vivo methods, and more established techniques, including, for example, immunocytochemistry and electrophysiological technologies. New trainees in neurosciences still need a sound footing in these older methods in order to apply a critical approach to their results.

Under the guidance of its founders, Alan Boulton and Glen Baker, the *Neuromethods* series has been a success since its first volume published through Humana Press in 1985. The series continues to flourish through many changes over the years. It is now published under the umbrella of Springer Protocols. While methods involving brain research have changed a lot since the series started, the publishing environment and technology have changed even more radically. *Neuromethods* has the distinct layout and style of the Springer Protocols program, designed specifically for readability and ease of reference in a laboratory setting.

The careful application of methods is potentially the most important step in the process of scientific inquiry. In the past, new methodologies led the way in developing new disciplines in the biological and medical sciences. For example, physiology emerged out of anatomy in the nineteenth century by harnessing new methods based on the newly discovered phenomenon of electricity. Nowadays, the relationships between disciplines and methods are more complex. Methods are now widely shared between disciplines and research areas. New developments in electronic publishing make it possible for scientists who encounter new methods to quickly find sources of information electronically. The design of individual volumes and chapters in this series takes this new access technology into account. Springer Protocols makes it possible to download single protocols separately. In addition, Springer makes its print-on-demand technology available globally. A print copy can therefore be acquired quickly and for a competitive price anywhere in the world.

Saskatoon, Canada *Wolfgang Walz*

Preface

During neuronal activity, neurotransmitters are released into the synaptic cleft and activate post- and presynaptic receptors to elicit actions such as neuronal firing. This process is known as neural transmission or neurotransmission. Importantly, this process has to be terminated, and therefore, the availability of neurotransmitter is the key determinant by which neurotransmission is regulated. Neurotransmitters either diffuse out of the closer vicinity to their sites of action or they are enzymatically inactivated (e.g., acetylcholine) or (for most other neurotransmitters) removed through reuptake by means of neurotransmitter transporters (NTTs). Thus, NTTs play a pivotal role in synaptic transmission. NTTs are important targets for clinically useful drugs (e.g., antidepressants and antiepileptic agents), but also for drugs of abuse (such as amphetamines or cocaine); they also play a role in the pathophysiology of diverse disorders of the central nervous system, e.g., attention deficit hyperactivity disorder (ADHD).

Monoamine transporters were the first which have been investigated almost five decades ago in landmark studies on "reuptake sites" by Georg Hertting in the laboratory of the late Nobel prize winner Julius Axelrod, a breakthrough at that time which was only made possible by the availability of radioactively labeled monoamines. In this early phase, the pharmacological and biochemical characterization relied on classical organ and tissue preparation techniques as well as on brain homogenates and platelets. The use of radioactively labeled NTT substrates and inhibitors enabled the characterization of substrate specificity, the identification of specific inhibitors and first insights in the cotransport of ions, the stoichiometric coupling of ion influx, and the substrate translocation including drug-induced reversal of the transport direction. These efforts also made possible, for example, the development of clinically important antidepressant drugs which act as selective monoamine transporter inhibitors such as the "selective serotonin reuptake inhibitors" (SSRIs). A major breakthrough was achieved some 25 years ago by the isolation of cDNAs that encode various transporters, starting with the cDNA of the transporters for GABA and norepinephrine. The tools of molecular biology enabled researchers to address the structural and functional properties of individual transporter proteins. By expressing transporter proteins in transfected cells, new functional features could be discovered such as channel-like properties with various types of currents. Furthermore, molecular biology techniques also made it possible to address the synthesis and intracellular delivery of the transporter proteins, their interaction with accessory proteins, and their regulated cycle between cell surface and endosomal compartments. The last great breakthrough with important and deep insights in the structure and function of NTTs was enabled by the development of highly efficient methods for the purification and crystallization of membrane proteins, including NTTs, as well as the availability of refined computer software for the modeling of structural and functional features of NTT proteins.

This volume *Neurotransmitter Transporters: Investigative Methods* within the *Neuromethods* series aims at providing an overview over cutting-edge methods and protocols used in neuroscience and neurological research. This book will be of interest to scientists, graduate students, and advanced undergraduates who seek an overview of methods and protocols in this field of neuroscience. It will also be of interest to physicians who are carrying out imaging and postmortem studies of neurotransmitter transporters in the human brain.

We would like to thank all contributors who gave us their unsolicited support to establish this book. Furthermore, we wish to express our gratitude to Wolfgang Walz, editor of the *Neuromethods* series (Humana Press), for the offered opportunity to produce this book and for his support in finalizing this exciting project.

While this volume was written, Georg Hertting passed away, one of the fathers of the field of neurotransmitter transport. Therefore, we wish to dedicate this volume to this great scientist.

Bonn, Germany — *Heinz Bönisch*
Vienna, Austria — *Harald H. Sitte*

Contents

Contributors

LUCIE BARTOVA • *Department of Psychiatry and Psychotherapy, Medical University Vienna, Vienna, Austria*

MARTIN BAUER • *Department of Psychiatry and Psychotherapy, Medical University Vienna, Vienna, Austria; Department of Clinical Pharmacology, Medical University Vienna, Vienna, Austria*

MICHAEL H. BAUMANN • *Medicinal Chemistry Section, IRP, NIDA, NIH, Baltimore, MD, USA*

BRUCE E. BLOUGH • *Center for Drug Discovery, Research Triangle Institute, Research Triangle Park, NC, USA*

HEINZ BÖNISCH • *Biomedical Center, Institute of Pharmacology and Toxicology, University of Bonn, Bonn, Germany*

SATHYAVATHI CHALLASIVAKANAKA • *Department of Biochemistry and Molecular Biology, University of North Dakota School of Medicine and Health Sciences, Grand Forks, ND, USA*

GIULIANO CIARIMBOLI • *Experimentelle Nephrologie, Medizinische Klinik D, Münster, Germany*

THOMAS COUROUSSÉ • *INSERM U1130, Paris, France; CNRS UMR 8246, Paris, France; UPMC Univ Paris 06, Sorbonne Universités, Paris, France; Université Paris Descartes, Ecole Doctorale Médicament Toxicologie Chimie Environnement, Paris, France*

LYNETTE C. DAWS • *Department of Physiology, University of Texas Health Science Center at San Antonio, San Antonio, TX, USA*

ANN M. DECKER • *Center for Drug Discovery, Research Triangle Institute, Research Triangle Park, NC, USA*

GERHARD F. ECKER • *Department of Pharmaceutical Chemistry, University of Vienna, Vienna, Austria*

BIRGIT EISENRAUCH • *Center for Physiology and Pharmacology, Institute of Pharmacology, Medical University of Vienna, Vienna, Austria*

JAMES D. FOSTER • *Department of Biochemistry and Molecular Biology, University of North Dakota School of Medicine and Health Sciences, Grand Forks, ND, USA*

MICHAEL FREISSMUTH • *Center for Physiology and Pharmacology, Institute of Pharmacology, Medical University of Vienna, Vienna, Austria*

LUKE R. GABRIEL • *Department of Psychiatry, Brudnick Neuropsychiatric Research Institute, University of Massachusetts Medical School, Worcester, MA, USA*

SOPHIE GAUTRON • *INSERM U1130, Paris, France; CNRS UMR 8246, Paris, France; UPMC Univ Paris 06, Sorbonne Universités, Paris, France*

TINA HOFMAIER • *Center for Physiology and Pharmacology, Institute of Pharmacology, Medical University Vienna, Vienna, Austria*

MARION HOLY • *Center for Physiology and Pharmacology, Institute of Pharmacology, Medical University of Vienna, Vienna, Austria*

ANDREAS JURIK • *Department of Pharmaceutical Chemistry, University of Vienna, Vienna, Austria*

DEEBA KHAN • *Center for Physiology and Pharmacology, Medical University of Vienna, Vienna, Austria*
OLEG KOVTUN • *Department of Chemistry, Vanderbilt University, Nashville, TN, USA*
FRANCK LOUIS • *INSERM U1130, Paris, France; CNRS UMR 8246, Paris, France; UPMC Univ Paris 06, Sorbonne Universités, Paris, France*
FELIX P. MAYER • *Center for Physiology and Pharmacology, Institute of Pharmacology, Medical University of Vienna, Vienna, Austria*
HALEY E. MELIKIAN • *Department of Psychiatry, Brudnick Neuropsychiatric Research Institute, University of Massachusetts Medical School, Worcester, MA, USA*
THERESE MONTGOMERY • *Center for Physiology and Pharmacology, Institute of Pharmacology, Medical University of Vienna, Vienna, Austria; School of Biomolecular and Biomedical Science, University College Dublin, Dublin, Ireland*
W. ANTHONY OWENS • *Department of Physiology, University of Texas Health Science Center at San Antonio, San Antonio, TX, USA*
JOHN S. PARTILLA • *Medicinal Chemistry Section, Molecular Targets and Medications Discovery Branch, Intramural Research Program; National Institute on Drug Abuse, National Institutes of Health, Baltimore, MD, USA*
DANIELA D. POLLAK • *Center for Physiology and Pharmacology, Medical University of Vienna, Vienna, Austria*
ANA POPOVIC • *Department of Psychiatry and Psychotherapy, Medical University Vienna, Vienna, Austria*
NICOLE PRASCHAK-RIEDER • *Department of Psychiatry and Psychotherapy, Medical University Vienna, Vienna, Austria*
DANIELLE E. RASTEDT • *Department of Biochemistry and Molecular Biology, University of North Dakota School of Medicine and Health Sciences, Grand Forks, ND, USA*
MARIANNE RONOVSKY • *Center for Physiology and Pharmacology, Medical University of Vienna, Vienna, Austria*
SANDRA J. ROSENTHAL • *Department of Chemistry, Vanderbilt University, Nashville, TN, USA; Department of Pharmacology, Vanderbilt University, Nashville, TN, USA; Department of Chemical and Biomolecular Engineering, Vanderbilt University, Nashville, TN, USA; Department of Physics and Astronomy, Vanderbilt University, Nashville, TN, USA; Vanderbilt Institute of Nanoscale Science and Engineering, Vanderbilt University, Nashville, TN, USA; Materials Science and Technology Division, Oak Ridge National Laboratory, Oak Ridge, TN, USA*
RICHARD B. ROTHMAN • *Medicinal Chemistry Section, Molecular Targets and Medications Discovery Branch, Intramural Research Program, National Institute on Drug Abuse, National Institutes of Health, Baltimore, MD, USA*
ULRICH SAUERZOPF • *Department of Psychiatry and Psychotherapy, Medical University Vienna, Vienna, Austria*
EBERHARD SCHLATTER • *Experimentelle Nephrologie, Medizinische Klinik D, Münster, Germany*
AMIR SEDDIK • *Department of Pharmaceutical Chemistry, University of Vienna, Vienna, Austria*
HARALD H. SITTE • *Center for Physiology and Pharmacology, Institute of Pharmacology, Medical University of Vienna, Vienna, Austria*
THOMAS STEINKELLNER • *Center for Physiology and Pharmacology, Medical University of Vienna, Vienna, Austria*

SONJA SUCIC • *Institute of Pharmacology, Center of Physiology and Pharmacology, Medical University of Vienna, Vienna, Austria*
GLENN M. TONEY • *Department of Physiology, University of Texas Health Science Center at San Antonio, San Antonio, TX, USA*
ROXANNE A. VAUGHAN • *Department of Biochemistry and Molecular Biology, University of North Dakota School of Medicine and Health Sciences, Grand Forks, ND, USA*
MATTHÄUS WILLEIT • *Department of Psychiatry and Psychotherapy, Medical University Vienna, Vienna, Austria*
SIJIA WU • *Department of Psychiatry, Brudnick Neuropsychiatric Research Institute, University of Massachusetts Medical School, Worcester, MA, USA*

Chapter 1

Classical Radioligand Uptake and Binding Methods in Transporter Research: An Emphasis on the Monoamine Neurotransmitter Transporters

Sonja Sucic and Heinz Bönisch

Abstract

Radioligand uptake and binding assays denote an invaluable tool in the field of neurotransmitter transporter research. Their benefits have been evident since the late 1950s, and they continue to contribute major insights into transporter function and structure to date. In the current chapter, we focus primarily on the family of monoamine (MA) neurotransmitter transporters (MATs), i.e., transporters (T) for norepinephrine [noradrenaline] (NET), dopamine (DAT), and serotonin [5-hydroxy tryptamine, 5-HT] (SERT), dysfunction of which has been linked to numerous neuropsychiatric disorders and substance abuse. Radiotracer assays have provided a major way of elucidating the mechanisms of action of not only endogenous substrates (e.g., NE, SER, and DA), but also many diverse substances such as antidepressants (e.g., imipramine and citalopram), psychostimulants (e.g., amphetamine and cocaine), toxins (e.g., conotoxins), or neurotoxins (e.g., 1-methyl-4-phenylpyridinium, MPP^+) that exert their action on MATs. In this chapter we describe the basic principles and experimental procedures of radiotracer assays commonly used in the studies of MATs.

Key words Radioligands, Uptake, Binding, Monoamine neurotransmitter transporters

1 Introduction

Noradrenaline (norepinephrine, NE) was the first monoamine to be recognized as a neurotransmitter [1]. Some 20 years later, the neurotransmitter functions of dopamine and serotonin (5-hydroxytryptamine, 5-HT) also became widely acknowledged. The availability of radioactive monoamines, beginning with tritiated (^{3}H) epinephrine in the late 1950s in the laboratory of the Nobel Prize laureate Julius Axelrod [2], made it possible to quantitatively study monoamine metabolism and tissue uptake. This led to the first demonstration of a cocaine-sensitive monoamine uptake system [3, 4]. Dengler et al. [5, 6] proposed that the uptake of ^{3}H-NE in sympathetically innervated tissues was mediated by a saturable membrane transport process. Iversen [7] was one of the first to describe the kinetic constants (K_m and V_{max}) of ^{3}H-NE

Heinz Bönisch and Harald H. Sitte (eds.), *Neurotransmitter Transporters: Investigative Methods*, Neuromethods, vol. 118, DOI 10.1007/978-1-4939-3765-3_1, © Springer Science+Business Media New York 2016

uptake into noradrenergic neurons of the isolated perfused rat heart by the NET (at the time still named "uptake 1"). By means of uptake studies using ^{3}H-monoamines, a vast array of substances and drugs were shown to interact with the monoamine transport systems, acting either as inhibitors (such as cocaine and diverse antidepressants) or as substrates (such as amphetamines), which lead to monoamine release by inducing reverse transport [8–10].

Prior to the cloning of MATs, these types of studies could only be performed in cell lines endogenously expressing the transporter(s) of interest. For example, the NET is natively expressed in rat PC12 pheochromocytoma cells and human SK-N-SH neuroblastoma cells [11] and the SERT in blood platelets, JAR human placental choriocarcinoma cells or rat RBL 2H3 basophilic leukemia cells. Using pools of clones from a SK-N-SH cell cDNA library transfected into COS-7 cells and identification of NET expressing transfectants by means of a radioactive NET substrate (^{125}I-m-iodobenzylguanidine), Susan Amara and coworkers [12] cloned the human NET as the first monoamine neurotransmitter transporter. Shortly thereafter, the DAT and SERT were cloned as well. In the 1980s, the first radioligand binding study on a MAT, using a tritiated antidepressant (^{3}H-imipramine), a potent inhibitor of serotonin transport, was published by Langer and coworkers [13]. The subsequent availability of additional tritiated, high-affinity MAT inhibitors (such as ^{3}H-nisoxetine for the NET or ^{3}H-citalopram for the SERT) allowed for further characterization of MATs by means of radioligand binding studies.

Meanwhile it became apparent that MATs belong to the solute carrier 6 (SLC6) gene family of transporters, which include not only the Na^{+}/Cl^{-}-dependent NET (SLC6A2), DAT (SLC6A3), and SERT (SLC6A4), but also the Na^{+}-dependent transporters for GABA and glycine [14]. Structural features common to all MATs are 12 transmembrane spanning domains and intracellular amino and carboxy termini. Moreover, multiple putative phosphorylation sites on intracellular domains support kinase-mediated regulation. The X-ray crystal structure of the *Drosophila melanogaster* DAT (dDAT), a eukaryotic transporter with approximately 50% sequence identity with its mammalian counterpart, became available in 2013 [15]. By means of radioligand binding and uptake experiments on dDAT mutants and determination of X-ray crystal structures of dDAT bound to substrates or inhibitors, the authors more recently identified molecular determinants that distinguish MAT substrates from inhibitors thereof [16]. The current chapter describes radioligand uptake and radioligand binding methods used in MAT research.

1.1 Basic Principles to Consider When Designing Radioligand Uptake and Binding Assays

Transporters are proteins located in biological membranes and their purpose is to translocate a solute from one side of the membrane to the other. This translocation process has been proposed to use an alternating access model of transport [17], which consists of four steps: (1) Reversible binding of the solute (ligand, L) to a binding site within the cavity of the outward open facing conformation state of the transporter (T), (2) if the ligand is a transported substrate (S), the binding induces a conformation change to the occluded state of the transporter and the translocation of the substrate through the membrane, (3) the dissociation (release) of the substrate at the inside in the inward open state, and finally (4) the reorientation to the unloaded transporter to the outward open state (see scheme below).

$$L + T \rightarrow [LT]; \text{the ligand } (L) \text{ can be a substrate } (S) \text{ or an inhibitor} \quad (\text{Step } 1);$$

If the ligand is a substrate (S, outside S_o or inside S_i):

$$S_o + T_{\text{outward facing}} \rightarrow [ST] \rightarrow S_i + T_{\text{inward facing}} \rightarrow T_{\text{outward facing}} \quad (\text{Steps } 2 \quad 4)$$

The binding step (step 1) is common for both types of ligands, translocated substrates and nontransported inhibitors, which cannot induce a conformation change in the transporter. The binding sites might not be identical for substrates and inhibitors, but overlapping. This simple model does not take into account Na^+ and Cl^- as co-substrates, i.e., the fact that all MATs are secondary active transporters which use the electrochemical gradient for Na^+ to facilitate the inward movement of the substrate across the membrane, even against a concentration gradient. In addition, the direction of transport can be reversed, e.g., via reversal of the Na^+-gradient or by substrates like indirectly acting amines (e.g., amphetamines) which—due to their inward transport—cause an outward transport of intraneuronal monoamines [9, 18].

1.1.1 Theoretical Considerations and Mathematical Models

Kinetics of Binding and Kinetics of Uptake (Transport)

The reversible binding of a ligand (radioligand) (L; which may be a substrate or an inhibitor) to its binding site within the transporter protein (T) leads to the formation of a ligand–transporter protein complex LT. The simplest assumption is that this process underlies the law of mass action (i.e., an interaction with a single noncooperative site), and thus follows the principles of Michaelis–Menten kinetics. Under this assumption, the formation of LT is determined by the rate constants for association (k_{on}) and dissociation (k_{off}). Equilibrium is obtained when the rate of formation (k_{on}; unit = M^{-1} min^{-1}) of the ligand–transporter complex (LT) equals the rate of dissociation (k_{off}; unit = min^{-1}) of this complex (Eq. 1):

$$[L][T]k_{on} = [LT]k_{off} \text{ or } L + T \underset{k_{on}}{\overset{k_{off}}{\rightleftarrows}} LT \quad (1)$$

At the equilibrium state, the equilibrium dissociation constant K_d (unit = M) can be derived from the ratio of k_{off} to k_{on} (Eq. 2) where [L], [T], and [LT] represent the concentrations of free ligand, free transporter protein, and bound ligand (or occupied transporter), respectively.

$$K_d = \frac{k_{off}}{k_{on}} = \frac{[L][T]}{[LT]} \tag{2}$$

Since the number of transporter proteins within a system is limited, the dependence of [LT] on [L] shows saturation and is described by a rectangular hyperbola defined by (Eq. 3):

$$[LT] = \frac{[LT]_{max}[L]}{K_d + [L]} \text{ or } B = \frac{B_{max}[L]}{K_d + [L]} \tag{3}$$

$[LT]_{max}$ or B_{max} is the total number of binding sites (transporters). The ligand concentration at which 50% of the binding sites are occupied describes the half-saturation concentration which is equal to the equilibrium dissociation constant K_d.

Saturable uptake of a ligand (radiolabeled substrate) by a transporter follows the same Michaelis–Menten kinetics as saturable binding given above, but the meanings are different, i.e., K_d becomes K_m and B_{max} becomes V_{max} (see below (Eq. 4)):

$$V = \frac{V_{max}[S]}{K_m + [S]} \tag{4}$$

The kinetic constants for an inhibitor (competitor) of binding or transport are given by the K_i value, which derives from the IC_{50}, i.e., the half-maximal inhibition of either binding or transport. The simplest (and most common) type of competition is the competitive inhibition where the competitor competes for the same (single) site at the transporter. From the IC_{50} values, the K_i values for a competitive inhibitor can be calculated according to the equation given by Cheng and Prusoff [19]:

$$K_i = \frac{IC_{50}}{1 + \left(\frac{[\text{Radioligand}]}{K_d}\right)} \text{ and (for transport) } K_i = \frac{IC_{50}}{1 + \left(\frac{[\text{substrate}]}{K_m}\right)} \tag{5}$$

1.1.2 Graphical Illustrations

The kinetics of association of binding of ^{3}H-desipramine to the NET and its dissociation from the transporter binding site is shown in Fig. 1.

Dissociation was induced by the addition (at equilibrium binding) of a high concentration (about 1000-fold K_i) of unlabeled competitor (nisoxetine). Insert (a) shows the linearization of the observed association rate (k_{ob} at the given radioligand concentration) and insert (b) the linearization of the dissociation (the slope equals k_{off}). The association rate constant (k_{on}) was calculated from

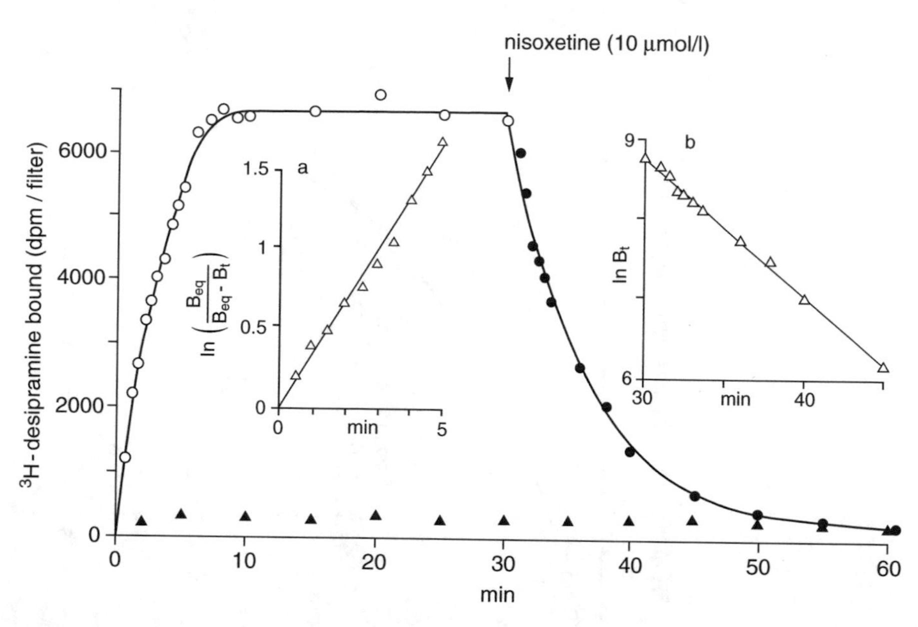

Fig. 1 Time-course of association and dissociation of binding of ^{3}H-desipramine (1.6 nM) to the NET. *Circles*: total binding, *filled triangles*: nonspecific binding in the presence of 10 µM nisoxetine, a high affinity and selective NET inhibitor (the data are taken with permission from ref. 20)

the equation $k_{on} = (k_{obs} - k_{off})/L$ and from $K_d = k_{off}/k_{on}$ (see Eq. 1) a K_d was obtained which was about identical with that from saturation binding (see Fig. 2).

Saturation of specific binding of ^{3}H-desipramine to the NET and linear increase with radioligand concentration of nonspecific binding. The insert shows the linearization of specific binding by means of a Scatchard plot (bound/free versus bound) from which K_d and B_{max} can be obtained from the slope $(=-1/K_d)$ and the intercept with the abscissa, respectively; taken from study of Bönisch and Harder [20] as in Fig. 1).

It should be noted that the transformation of the experimental data into a linear form and the subsequent analysis by linear regression, using plots previously described by Lineweaver–Burk, Eadie–Hofstee or Scatchard (or Rosenthal), should only be used as an illustration of results (e.g., to show how a treatment changes K_d or B_{max}). Currently, nonlinear regression programs such as GraphPad Prism (GraphPad Software, Inc. La Jolla, CA, USA) should be used, since such software enables not only the calculation of the requested kinetic constants (even for more complex forms of binding with Hill coefficients deviating from unity), but also drawing of diverse types of graphs and statistic comparisons (see Sect. 2.3).

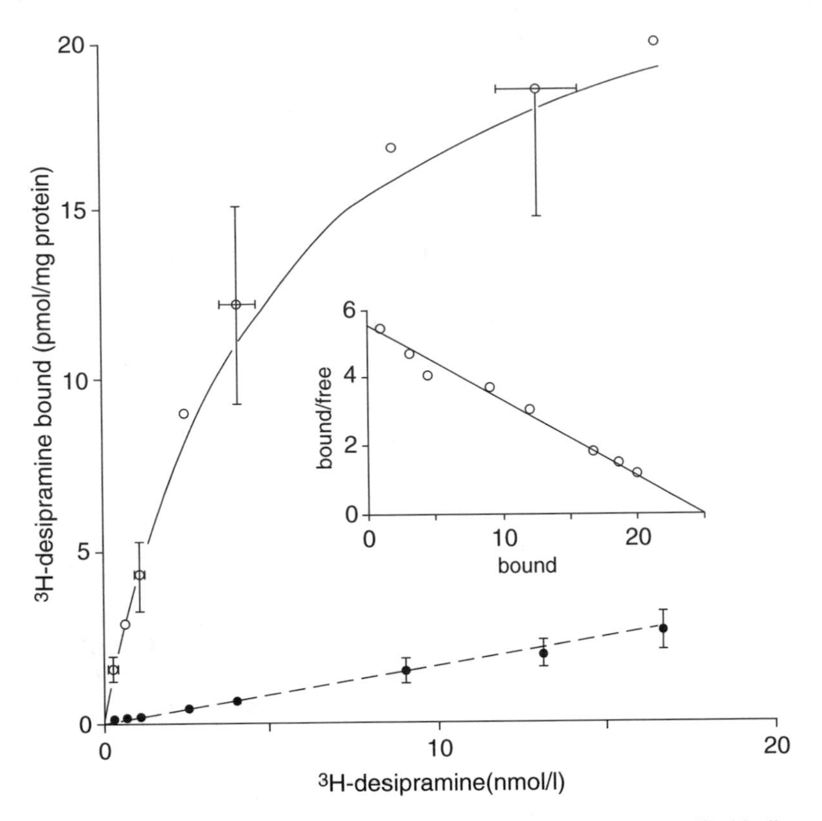

Fig. 2 Specific binding of ^{3}H-desipramine (*open circles*) and nonspecific binding (*filled circles*) in the presence of nisoxetine (10 μM, i.e., about 1000-fold its K_i value). The *insert* shows the linearization of specific binding according to Scatchard (the data are taken with permission from ref. 20)

1.1.3 Comments

(a) Binding studies

1. Only ^{3}H or ^{125}I labeled compounds with high specific activity and selectivity for the target protein are useful for radioligand binding studies.
2. The radioligand should exhibit high affinity to only the binding site of interest.
3. The K_d for the binding site should be <100 nM (better <10 nM). If the K_d is >10 pM, the dissociation rate might be very slow and it will be difficult to achieve equilibrium. If the K_d is higher than about 100 nM, the dissociation will be fast and a lot of the bound radioligand will disappears during the washing of the filters; this will lead, at least, to a pronounced underestimation of B_{max}.
4. The radioligand should exhibit low nonspecific binding to the biological material and/or the filters. Thus, for a novel radioligand at first binding to various filters and nonspecific binding should be determined.

5. Nonspecific binding of the radioligand to glass fiber filters can be reduced by presoaking the filters with polyethyleneimine (PEI 1–5%).
6. Usually glass fiber filters (e.g., GF/B or GF/F) are used to separate bound radioligand from its free form; separation of free ligand from a solubilized transporter protein can be performed using Sephadex G50 molecular sieve columns [21].
7. The unlabeled competitor used to determine nonspecific binding should be used at a concentration more than 100-fold its K_i and it should exhibit high affinity to the binding site and low nonspecific binding; the latter often increases with the hydrophobicity of the radioligand.
8. Nonspecific binding at the highest concentration of radioligand should be less than 50% of total binding.
9. Binding must be performed at equilibrium, and if this is unknown for the radioligand and its binding site, it has to be determined experimentally with a low concentration of the radioligand relative to its K_d.
10. A low temperature (e.g., 4 °C) is often of advantage because low temperature decreases degradation of biological material and often also nonspecific binding (to for example filters).
11. The highest concentration of radioligand should be at least ten times the K_d, which means that the occupancy of the binding site exceeds 90%.
12. To avoid depletion, binding of the radioligand should not use up more than 10% of free radioligand, otherwise the amount of transporter protein must be reduced or the sample volume increased.
13. Affinity constants (K_i or K_d values) do not tell us whether the ligand is a substrate or an inhibitor. This information can be obtained only from transport studies by demonstrating either uptake of the radiolabeled ligand and/or induction of carrier-mediated efflux.

(b) Transport (uptake) studies

1. Can only be performed with biological system existing as a closed compartment such as intact cells (isolated cells, tissue pieces, perfused organs), synaptosomes or plasma membrane vesicles (for the latter, see ref. 22).
2. Should be carried out at a physiological temperature (37 °C) or at least at room temperature since transport is—at least for MATs—a secondary active process driven by the sodium gradient which is maintained by the sodium/potassium ATPase.
3. The assay buffer for uptake studies should be a physiological salt solution containing glucose (for maintenance of metabolic

energy). In addition, a possible degradation of catecholamine substrates by the enzymes MAO or COMT should be prevented by the addition of an inhibitor of these enzymes. HEK293 cells which are often used for transfection with MAT cDNAs have been reported to express COMT [23].

4. In contrast to radioligand binding experiments, transport studies should measure (if possible) initial rates of uptake, i.e., within a time scale when the intracellular accumulation of the radioactive substrate still shows a linear increase with time. Otherwise, V_{max} will be underestimated. Thus, a time-course of uptake should be determined before measuring saturation kinetics.
5. Uptake studies—in contrast to binding studies—can also be done using ^{14}C-labeled substrates (e.g., ref. 24).
6. To determine the effect of a noncompetitive inhibitor on transport and binding or to estimate the turnover number of a transporter from V_{max}/B_{max}, both, uptake and binding should be carried out in intact cells (e.g., ref. 25).
7. As a rule, substrates are less bulky molecules compared to uptake inhibitors and substrates show relatively high k_{off} rates and thus relatively low affinity to the transporter (K_i or K_m values higher than 100 nM). This is conceivable since substrates should rapidly dissociate from the transporter after being translocated.

For further literature on uptake and binding methods, the reader is referred to the following reviews or books and chapters therein [26–32]. In addition, protocols are also available online (see e.g.: https://pdspdb.unc.edu/pdspWeb/?site=assays).

1.2 Key Parameters in MAT Research Determined by Means of Radioligand Assays

The substrate spectrum of monoamine transporters (MATs) ranges from endogenous catecholamines, noradrenaline [7, 33], adrenaline [34], and dopamine [35], to serotonin [36], Parkinsonism-inducing neurotoxins such as 1-methyl-4-phenylpyridinium (MPP^+) [37] and the psychostimulant amphetamine [9]. Monoamine transporters are the site of action of different classes of antidepressants (transport inhibitors), i.e., the tricyclic antidepressants, such as desipramine and imipramine [38, 39], the non-tricyclic antidepressants, such as reboxetine and paroxetine [40, 41] and psychostimulant drugs, such as cocaine [33]. As described in detail in Sect. 1.1, the substrate uptake mechanism by MATs can be described by a simple kinetic model, which allows the transporter to exist in three different states in the transport cycle: (1) the binding of free extracellular substrate to the outward facing substrate recognition site on the transporter, (2) the translocation of the substrate and its subsequent dissociation inside the cell and (3) the reorientation of the unloaded transporter to a state in which it is able to begin another transport cycle [17, 42].

Substrate binding to MATs and their subsequent translocation into the cell is dependent on the co-substrates Na^+ and Cl^- ions that are the driving forces of the transport process [43]. This model can be used to describe transporter function by determining kinetic parameters from pharmacological studies. Substrate binding to the transporter (state 1 of the kinetic model of transport) can be determined from the K_i value of a substrate for inhibition of ^{3}H-inhibitor binding. The K_i value obtained from binding assays is inversely related to the substrate binding affinity because it depends only on substrate competition for the interaction of inhibitor with the transporter. The K_m value determined from [^{3}H]substrate uptake assays is a more complex parameter, since it is influenced not only by the binding affinity ($1/K_i$) of a substrate for the transporter, but also by other steps in the transport cycle. The K_m value is inversely related to the apparent affinity of a substrate for the transporter. The apparent affinities ($1/K_m$) of substrates have previously been shown to be lower than the binding affinities ($1/K_i$). For instance, the binding affinity of norepinephrine was 50-fold greater than its apparent affinity for hNET (Bönisch and Harder [20]), which is consistent with the kinetic model of substrate transport described above. In contrast to the studies in intact cells, the binding affinity ($1/K_i$) of norepinephrine determined from ^{3}H-desipramine binding in membrane preparations was approximately 70-fold less than its apparent affinity determined in uptake studies ($1/K_m$) [44]. High affinity binding of substrates has only been observed in intact cells, indicating that a different conformation of the transporters may exist in intact cells compared with dispersed cell membranes. This difference can be attributed to an ionic disequilibrium or to a missing interaction of the transporters with intracellular MAT modifiers, i.e., interleukins such as IL1beta [45] and IL6 [46], cAMP [47], protein kinase C [48], which are only present in intact cells, but absent in isolated membrane preparations. In light of these reports, the data from studies carried out in intact cells may be more relevant in terms of mirroring the physiological state, compared to experiments on isolated membrane preparations.

Another key aspect reflecting MAT function is the transporter turnover number, i.e., the number of transport cycles per unit time, which is reflective of substrate translocation mechanisms and/or conformational changes associated with substrate transport. Transporter turnover can be calculated from the ratio V_{max}/B_{max} [20]. The V_{max} of ^{3}H-substrate uptake represents the maximal initial rate of substrate uptake, and it depends on the capacity of the transporter for substrate uptake and the amount of transporter protein at the cell membrane. The B_{max} of ^{3}H-inhibitor binding is the maximal binding, reflecting transporter surface expression. It is crucial to note that V_{max} of binding of the potent and selective NET inhibitor nisoxetine to the NET in intact cells has previously been shown to be highly correlated with transporter expression data from cell surface biotinylation experiments [49]. When examining interactions of transporters with inhibitors, the K_i or K_d values determined from inhibition of ^{3}H-inhibitor binding are inversely related to their binding affinities, because they depend only on inhibitor competition for the interaction of the given inhibitor with the transporter. The K_i values determined from inhibition of ^{3}H-substrate uptake are inversely related to their "apparent" affinities, because they depend on inhibitor competition for the interaction of substrate with the transporter, which is related to the mechanisms involved in substrate uptake (i.e., states 1–3 of the kinetic model of transport, described above). These parameters have been extensively investigated for the human (h) NET (hNET); the binding affinities of inhibitors ($1/K_d$) were reported to be equal to their apparent affinities for uptake inhibition ($1/K_i$) [50–53]. This finding suggested that inhibitors compete for the same (or strongly overlapping) sites on the NET, regardless of whether they are inhibiting substrate uptake or inhibitor binding, such that the binding sites for substrates and inhibitors on the transporter must be at least overlapping.

1.3 Radioligand Uptake and Binding Are Valuable Assays to Study the Nature of Interactions of MATs with their Ligands

Radiotracer-based assays have to a vast effect helped elucidate the mechanisms of action of many MAT ligands. For instance, transport studies in PC12 cells revealed that the psychostimulant ^{3}H-amphetamine enters the cells in a carrier-mediated manner, via the NET [10]. Apart from studies on wild type MATs, the use of chimeras of NET, DAT, and SERT has been a helpful tool in delineating the functional roles of broad transporter domains involved in the interactions with substrates, ions or inhibitors [54–56]. They were also beneficial in conferring targets for the early site-directed mutagenesis studies [57, 58]. The main disadvantage of the chimeric approach was that a relatively high percentage of chimeras yielded nonfunctional proteins, but site-directed mutagenesis has been widely employed to explore structure–function relationships of MATs. MATs are proteins of over 600 amino acid residues and many mutants of these transporters have been studies since the 1990s (for review, see ref. 59). When interpreting mutagenesis data, one must be aware that the effects of mutations may arise from indirect conformational changes in the protein. Moreover, the majority of mutagenesis studies report only the apparent affinities of ligands, determined from inhibition of ^{3}H-substrate uptake. The "real," binding affinities determined from inhibition of ^{3}H-inhibitor binding are often not available, making it difficult to distinguish between the binding sites for substrates and inhibitors on MATs. Radioligand-based assays have also been useful in elucidating the basis underlying some disease states, related to naturally occurring transporter variants, e.g., mutations in the hNET gene have been extensively examined in the late 1990s [53, 60–62].

2 Materials

2.1 Cell Culture

1. Cell lines used for transfection with MAT cDNAs: HEK293, HeLa, COS-7, CAD, JAR (American Type Culture Collection, Bethesda, MD, USA).
2. Dulbecco's modified Eagle's medium (DMEM) or RPMI (Gibco, Invitrogen Life Science, Bethesda, MD).
3. Fetal bovine serum (FBS) (Carlsbad, CA).
4. 0.05 % trypsin–EDTA (Sigma-Aldrich).
5. T25/T75 flasks for maintaining the cell lines; for uptake and binding assays on intact cells, the cells are seeded onto 24-, 48-, or 96-well culture plates (BD Biosciences, Falcon; Techno Plastic Products (TPP) cultureware).
6. Penicillin (10,000 U/mL) and streptomycin (10 mg/mL) solutions are frozen at −20 °C (Gibco, Invitrogen Life Science); 5 mL is added to 0.5 L of DMEM complete culture medium.
7. Poly-D-lysine solution (0.1 mg/mL), prepared in sterile H_2O, is used to coat the surface of the culture plates for firm attachment of the cells.
8. Cell culture incubator, 37 °C, 5 % CO_2.
9. Vacuum pump for cell washing.

2.2 Radioligands and Some Drugs Typically Used in Uptake and Binding Assays on MATs

Unlabeled substances (drugs, substrates, inhibitors) used in MAT research are available from Sigma-Aldrich or Sigma/RBI (St. Louis, MO, USA) (refer to: www.sigmaaldrich.com/technical-documents/articles/biology/rbi-handbook/transporters/biogenic-amine-transporters.html) or from Tocris Bioscience (Bristol, UK) (see: http://www.tocris.com/pharmacological-Browser.php?ItemId=187632#.ViOwnm6BQg4).

1-Methyl-4-phenylpyridinium ion (MPP^+) can be purchased from Sigma/RBI (Natick, MA). The supplements required in assay buffer, used to study uptake in intact cells, are available from the following companies: ascorbic acid, to prevent monoamine oxidation (Sigma-Aldrich), pargyline hydrochloride, to inhibit monoamine oxidase (MAO; Sigma-Aldrich) and U-0521 (3′,4′-dihydroxy-2-methylpropiophenone), to inhibit catecholamine oxidase (COMT) (Pharmacia and Upjohn, Kalamazoo, MI). Tritiated or ^{125}I-labeled compounds used in MAT research are available from Perkin Elmer Life and Analytical Sciences (Waltham, MA, USA) (homepage: http://www.perkinelmer.com) or from ARC (American Radiolabeled Chemicals, Inc, Saint Louis, MO, USA; www.arc-inc.com).

(A) Labeled substrates (to be used on the given MAT)

1. ^{3}H-5-hydroxytryptamine (^{3}H-5-HT, serotonin), from PerkinElmer or ARC (for SERT).
2. ^{3}H-1-methyl-4-phenylpyridinium ion (^{3}H-MPP$^+$) supplied by ARC (for NET, DAT [and SERT]).
3. ^{3}H-norepinephrine from PerkinElmer or ARC (for NET).
4. ^{3}H-dopamine from PerkinElmer or ARC (for DAT and NET).

(B) Labeled inhibitors (to be used on the given MAT)

1. ^{3}H-WIN 35,428 ((−)-2-β-Carbomethoxy-3-β-(4-fluorophenyl) tropane, β-CFT) from PerkinElmer or ARC (for DAT).
2. ^{3}H-GBR 12935 (1-[2-(Diphenylmethoxy)ethyl]-4-(3-phenylpropyl)piperazine) from ARC (for DAT).
3. ^{125}I-RTI-55 (2-β-carbomethoxy-3-β-[4′-iodophenyl]tropane; β-CIT) from ARC (for SERT, DAT [and NET]).
4. ^{3}H-nisoxetine, from PerkinElmer or ARC (for NET).
5. ^{3}H-desipramine, from PerkinElmer or ARC (for NET).
6. ^{3}H-imipramine, from PerkinElmer or ARC (for SERT).
7. ^{3}H-citalopram, from PerkinElmer or ARC (for SERT).
8. ^{3}H-escitalopram, from PerkinElmer (for SERT).

Table 1 displays basic kinetic parameters of selected ligands of MATs [50–53, 63–68]. The pharmacological profiles of other MAT ligands have also been well described in the literature (e.g., ref. 69).

2.3 Equipment, Software, and Data Analysis

1. Skatron semi-automatic cell harvester (Lier, Norway) for binding assay filtration using isolated membrane preparations.
2. Packard 2500 TR Liquid Scintillation Analyzer for measuring ^{3}H-radioactivity. For studies using ^{125}I-labeled compounds, a gamma scintillation counter is required to measure ^{125}I content (such as WIZARD2; http://www.perkinelmer.com/Catalog/Family/ID/WIZARD2%20Automatic%20Gamma%20Counter).
3. Bio-Tek ELx405R Automated Microplate Washer, only for assays performed in a 96-well format, as described by Lynagh and coworkers [70].
4. GraphPad Prism (Prism 5 or 6, GraphPad Software, San Diego, CA) for analysis of uptake and binding data. Specific uptake or binding is calculated by subtracting the nonspecific values (determined in the presence of a MAT inhibitor) from total counts. K_m and V_{max} values for specific uptake of ^{3}H-substrates and K_d and B_{max} values for specific [^{3}H]inhibitor binding are calculated by nonlinear regression analysis of the data according to a hyperbolic model. IC_{50} values for inhibition of uptake or

Table 1

Basic kinetic parameters of selected MAT ligands, determined by means of radiotracer uptake and binding assays

MAT	Substrates				Inhibitors		K_i values	Ref.
hSERT	K_m (5-HT)	V_{max} (5-HT)	K_m (MPP^+)	V_{max} (MPP^+)	K_D (imipramine)	B_{max} (imipramine)	Cocaine: 740 nM	[63]
					(By isolated plasma membrane binding)			[64]
							Methylphenidate: 132 μM	[65]
	1.8 μM	151 pmol/10^6 cells/min	24.2 μM	57 pmol/10^6 cells/min	4.5 nM	15 pmol/mg protein		[66]
							Amphetamine: 38.5 μM	
hNET	K_m (NA)	V_{max} (NA)	K_m (MPP^+)	V_{max} (MPP^+)	K_D (nisoxetine)	B_{max} (nisoxetine)	Cocaine: 480 nM	
					(By whole cell binding)			
	1.9 μM	105 pmol/10^6 cells/min	0.78 μM	79.8 pmol/mg protein/min	4.4 nM	0.97 pmol/mg protein	Methylphenidate: 0.1 μM	[50]
								[51]
	K_m (DA)	V_{max} (DA)			K_D (desipramine)	B_{max} (desipramine)	Amphetamine: 0.07 μM	[52]
					(By isolated plasma membrane binding)			[53]
	0.69 μM	56.4 pmol/10^6 cells/min			5.7 nM	1.8 pmol/mg protein	Nisoxetine: 4.9 nM	[64]
hDAT	K_m (DA)	V_{max} (DA)	K_m (MPP^+)	V_{max} (MPP^+)	K_D (CFT)	B_{max} (CFT)	Cocaine: 230 nM	
					(By isolated plasma membrane binding)			[64]
	2.2 μM	8.2 pmol/10^6 cells/min	0.017 μM	31.7 pmol/10^6 cells/min	35 nM	3.75 pmol/mg protein	Methylphenidate: 60 nM	[66]
								[67]
					12 nM (*with 10 μM Zn2+*)	6.75 pmol/mg protein (*with 10 μM Zn2+*)	Amphetamine: 0.64 μM	[68]

The given kinetic parameters were obtained from experiments carried out in transiently transfected COS-7 or HEK293 cells, or membranes thereof as indicated. The listed K_i values were determined from inhibition of substrate uptake by the indicated drugs binding by various drugs are calculated by nonlinear regression analysis of percentage inhibition data versus log drug concentration according to a sigmoidal model. The IC_{50} values are used to calculate the K_i values, assuming competitive inhibition [19].

3 Methods

3.1 Substrate Uptake Assays

1. One day prior to performing uptake assays, the cells (e.g., COS-7, HEK293, CAD, HeLa) transiently or stably expressing MATs of interest are seeded onto multiple well (24-, 48- or 96-well) plates, at a density of $0.4–2 \times 10^5$ cells per well, according to the experimental design; for details of transfection methods, see chapter of Steinkellner et al. Some cell lines endogenously express MATs, e.g., SERT-expressing choriocarcinoma JAR cell line. In addition, cell lines stably expressing MATs are now commercially available from some companies (e.g., www.genescript.com or www.abcam.com). It is essential to coat the plates with 0.1 mg/ml poly-D-lysine (PDL) solution prior to seeding the cells for uptake assays, to avoid cell detachment during the multiple washing steps during the experiment (i.e., add the PDL solution onto each well, covering the entire well surface. After a minimum of 30 min, the PDL solution can be aspirated and the wells washed twice with sterile Milli-Q water).
2. On the day of experiment, culture medium is aspirated from the wells, followed by gentle washing with Krebs-HEPES buffer at 25 °C [composition (in mM): NaCl 125, KCl 4.8, $MgSO_4$ 1.2, KH_2PO_4 1.2, $CaCl_2$ 1.3, HEPES 25, D(+)-glucose 5.55, ascorbic acid 1.02, pH 7.4] containing 0.1% bovine serum albumin (BSA; Sigma-Aldrich, Catalog No. A7906). For experiments using catecholamines (NE and DA), the KRH buffer should be supplemented with 10 μM U-0521 (3′,4′-dihydroxy-2-methylpropiophenone), to inhibit catechol *O*-methyltransferase [COMT; *S*-adenosyl-L-methionine:catechol *O*-methyltransferase; EC 2.1.1.6]) and with 1 μM pargyline (to inhibit monoamine oxidase [MAO; amine:oxygen oxidoreductase (deaminating) (flavin containing); EC 1.4.3.4]). In the case of ^{3}H-5-HT or ^{3}H-MPP^+ being used as substrates in uptake assays, it is not necessary to include U-0521 or pargyline in the KRH buffer.
3. The cells are preincubated for 10 min at 25 °C with Krebs-HEPES buffer in the absence or presence of the specific inhibitor (i.e., 10 μM paroxetine, mazindole, or nisoxetine, for SERT, DAT, or NET, respectively) to determine nonspecific uptake, and 0.1–60 μM ^{3}H-5-HT, ^{3}H-dopamine or

^{3}H-norepinephrine for precisely 1–2 min (note that a longer incubation time of 2 min is typically used for measuring uptake by the NET and DAT).

4. The uptake is rapidly terminated by washing the cells with ice-cold Krebs-HEPES buffer (kept at 0 °C). The washing can be done by carefully flipping the cold buffer solutions (prefilled into the appropriate multi-well plates and kept on ice) onto the cells. The procedure should be repeated three times. Alternatively, the uptake assay can be performed at 10 s intervals between wells, and the wells are then washed with cold buffer at the exact time point to maintain the exact incubation time.
5. The cells are finally lysed in 1% SDS (or 0.1% Triton X-100) and assayed for ^{3}H and protein content or cell count number. Specific activity must be calculated for each experiment.

3.2 Inhibitor Binding Assays

3.2.1 Binding Assays Using Isolated Membrane Preparations

1. Membranes are prepared from cells expressing the MAT of interest, with expression levels typically varying in the range of 3–30 pmol/mg protein, depending on transfection efficiency. All steps should be done on ice: 48 h after transfection, the medium is removed, and the cell layer washed thrice with ice-cold PBS. The cells are mechanically detached in PBS (cell scrapers can be used to collect the cells) and harvested by centrifugation (15,000 rpm for 10 min). The cell pellet is resuspended in 0.5 ml of hypotonic buffer (25 mM HEPES, 2 mM $MgCl_2$, and 1 mM EDTA, pH 7.3) in the presence of a mixture of protease inhibitors (e.g., SIGMAFAST™ Protease Inhibitor Cocktail Tablets, Sigma-Aldrich). It is then frozen in liquid nitrogen, followed by two rapid freeze-thaw cycles and sonication (thrice for 10 s). Cell membranes are collected by centrifugation at 40,000 × g for 15 min. The membranes are resuspended in the same buffer, to a protein concentration of 5 mg/mL and frozen in liquid nitrogen.
2. For SERT, ^{3}H-imipramine or ^{3}H-citalopram equilibrium binding is performed in duplicate incubations in an assay volume of 0.2–0.5 mL (adjusted to avoid radioligand depletion).
3. In competition binding experiments, membrane preparations (5–20 μg/assay) are incubated with the radioligand (e.g., 2 nM ^{3}H-imipramine, which is the K_d value of ^{3}H-imipramine for SERT) and the indicated concentrations of the competing drugs in assay buffer (20 mM Tris-HCl, 1 mM EDTA, 2 mM $MgCl_2$, 3 mM KCl, and 120 mM NaCl, pH adjusted to 7.4). Nonspecific binding is determined in the presence of 10 μM paroxetine. Saturation experiments are done with serial dilutions of ^{3}H-imipramine ranging from 0.1 to 50 nM.
4. For experiments on DAT-expressing membranes, the assay buffer should also be supplemented with 10 μM $ZnCl_2$ and be devoid of

EDTA, since Zn^{2+} promotes the outward-facing conformation that is required for high-affinity binding of inhibitors [71].

5. Equilibrium binding of ^{3}H-WIN35,428 to DAT is performed in an assay volume of 0.1 ml containing 10 nM ^{3}H-WIN35,428 (for competition binding experiments). In saturation experiments, the concentrations of ^{3}H-WIN35428 should range from 1 to 100 nM. Nonspecific binding is determined in parallel in the presence of 10 μM methylphenidate.
6. Binding of ^{3}H-desipramine to the NET on plasma membranes has been described by Bönisch and Harder [20]. In brief, the membranes (25 μg) are incubated at 25 °C for 30 min with ^{3}H-desipramine (1 nM) in buffer (135 mM NaCl, 10 mM Tris-HCl (pH 7.4), 5 mM KCl, 1 mM $MgSO_4$, 1 mM dithiothreitol) in a total assay volume of 250 μl. Reaction is stopped by the addition of ice-cold buffer (1 ml) and the mixture filtered through polyethyleneimine-pretreated GF/F glass fiber filters (for details, see below), which are then washed twice with 1 mL buffer at 0 °C. Nonspecific binding is determined in the presence of 10 μM nisoxetine.
7. Binding is normally allowed to proceed for 15–60 min at 20 °C and is terminated by rapid filtration onto GF/A glass microfiber filters (Whatman International Ltd, Maidstone, UK) presoaked in 0.5% polyethyleneimine (Sigma-Aldrich) to reduce nonspecific binding to the filter. The radioactivity trapped onto the filters is measured by liquid scintillation counting.

3.2.2 Binding Assays in Intact Cells

1. For measurement of inhibitor binding to intact cells, culture medium is removed from the cells firmly attached to the wells of for example 24-well plates and the cells are washed carefully thrice with 1 ml Krebs-HEPES buffer at 0 °C.
2. The cells are incubated for 60 min at 0 °C with Krebs-HEPES buffer containing 0.1% BSA and ^{3}H-nisoxetine (0.5–60 nM) in the absence and presence of 200 μM dopamine (to determine total and nonspecific binding, respectively) and other drugs, as necessary. In ion (e.g., Na^+) replacement studies, the NaCl concentration of the Krebs-HEPES buffer must be adjusted, such that 150 mM NaCl is present in the controls, and for the other wells in the experiment, up to 140 mM of this NaCl is replaced by equiosmolar concentrations of LiCl or *N*-methyl-D-glucamine (NMDG).
3. The cells are finally lysed in 1% SDS (or 0.1% Triton X-100) and assayed for ^{3}H and protein content or cell count number. Specific activity should be calculated for each experiment, by taking 100 μl samples of the incubation solutions.

Notes

1. Binding to intact cells may also be carried out at ambient temperature or even at 37 °C (e.g., ref. 25).
2. An uptake assay using 96-multiwells has recently been published [70].
3. Scintillation proximity assays (SPA) can be used in transporter research. Information on SPA can be found on the Perkin Elmer homepage, under: www.perkinelmer.com/catalog/category/id/spa.
4. In uptake and binding assays involving methanethiosulphonate (MTS) reagents, the Krebs-HEPES buffer should be slightly modified by replacing 1 mM ascorbic acid by 40 μM ethylenediaminetetraacetate (Na_2EDTA) and 0.1 mM ascorbic acid. This buffer is used for washing the cells prior to the assays and for removing the MTS reagents and/or drugs as required. The experimental details relevant to the use of MTS reagents in substituted cysteine accessibility method (SCAM) studies have been described in detail by Karlin and Akabas [72] and Javitch [73]. The SCAM method has been successfully applied to MATs [74–76]. In preliminary tests to determine transporter sensitivity to MTS reagents, the cells are preincubated with the vehicle solution (25 μl water in 1 ml Krebs-HEPES buffer; control), 1 mM 2-(trimethylammonium)ethyl-MTS (MTSET) (25 μl 40 mM stock solution in 1 ml Krebs-HEPES), or 2.5 mM MTSEA (2-aminoethyl-MTS) (25 μl 100 mM stock solution in 1 ml Krebs-HEPES buffer) for 5 min at 25 °C. The cells are then washed thrice with the same buffer at 25 °C, after which the uptake assays can be initiated. For some of the cysteine mutants used in these studies, the cells should be preincubated with 1.25 mM MTSEA (25 μl 50 mM stock solution in 1 ml Krebs-HEPES) at 25 °C for 10 min, in order to establish experimental conditions required for subsequent concentration- and time-dependence experiments. In concentration-dependence assays, the cells are preincubated with 0.078–1.25 mM MTSEA for 10 min at 25 °C. In time-dependence experiments, the cells are exposed to 1.25 mM MTSEA for 0–10 min at 25 °C, in the absence or presence of various inhibitors and substrates, to assess the ability of various ligands to protect against MTS reagent reactivity. The concentration of a ligand tested for protection should be 100-fold higher than its apparent affinity.

References

1. Von Euler US (1946) The presence of a sympathomimetic substance in extracts of mammalian heart. J Physiol 105:38–44
2. Axelrod J, Weil-Malherbe H, Tomchick R (1959) The physiological disposition of H3-epinephrine and its metabolite metanephrine. J Pharmacol Exp Ther 127:251–256
3. Whitby LG, Hertting G, Axelrod J (1960) Effect of cocaine on the disposition of noradrenaline labelled with tritium. Nature 187:604–605
4. Herting G, Axelrod J, Whitby LG (1961) Effect of drugs on the uptake and metabolism of H3-norepinephrine. J Pharmacol Exp Ther 134:146–153
5. Dengler HJ, Spiegel HE, Titus EO (1961) Uptake of tritium-labeled norepinephrine in brain and other tissues of cat in vitro. Science 133(3458):1072–1073
6. Dengler HJ, Michaelson IA, Spiegel HE, Titus E (1962) The uptake of labelled norepinephrine by isolated brain and other tissues of the cat. Int J Neuropharmacol 1:23–38
7. Iversen LL (1963) The uptake of norepinephrine by the isolated perfused rat heart. Br J Pharmacol 21:523–537
8. Iversen LL (1967) The uptake and storage of noradrenaline in sympathetic nerves. Cambridge University Press, Cambridge, UK
9. Langeloh A, Bönisch H, Trendelenburg U (1987) The mechanism of the ^{3}H-noradrenaline releasing effect of various substrates of uptake1: multifactorial induction of outward transport. Naunyn Schmiedebergs Arch Pharmacol 336:602–610
10. Bönisch H (1984) The transport of (+)-amphetamine by the neuronal noradrenaline carrier. Naunyn Schmiedebergs Arch Pharmacol 327:267–272
11. Bönisch H, Brüss M (2006) The norepinephrine transporter in physiology and disease. Handb Exp Pharmacol 175:485–524
12. Pacholczyk T, Blakely RD, Amara SG (1991) Expression cloning of a cocaine- and antidepressant-sensitive human noradrenaline transporter. Nature 350(6316):350–354
13. Raisman R, Briley MS, Langer SZ (1980) Specific tricyclic antidepressant binding sites in rat brain characterised by high-affinity ^{3}H-imipramine binding. Eur J Pharmacol 61:373–380
14. Pramod AB, Foster Carvelli JL, Henry LK (2013) SLC6 transporters: structure, function, regulation, disease association and therapeutics. Mol Aspects Med 34:197–219
15. Penmatsa A, Wang KH, Gouaux E (2013) X-ray structure of dopamine transporter elucidates antidepressant mechanism. Nature 503(7474):85–90
16. Wang KH, Penmatsa A, Gouaux E (2015) Neurotransmitter and psychostimulant recognition by the dopamine transporter. Nature 521(7552):322–327
17. Rudnick G, Krämer R, Blakely RD, Murphy DL, Verrey F (2014) The SLC6 transporters: perspectives on structure, functions, regulation, and models for transporter dysfunction. Pflugers Arch 466:25–42
18. Trendelenburg U (1990) Carrier-mediated outward transport of noradrenaline from adrenergic varicosities. Pol J Pharmacol Pharm 42:515–520
19. Cheng Y, Prusoff W (1973) Relationship between the inhibition constant (Ki) and the concentration of inhibitor which causes 50 per cent inhibition (I50) of an enzymatic reaction. Biochem Pharmacol 22:3099–3108
20. Bönisch H, Harder R (1986) Binding of ^{3}H-desipramine to the neuronal noradrenaline carrier of rat phaeochromocytoma cells (PC-12 cells). Naunyn Schmiedebergs Arch Pharmacol 334:403–411
21. Schömig E, Bönisch H (1986) Solubilization and characterization of the ^{3}H-desipramine binding site of rat phaeochromocytoma cells (PC12-cells). Naunyn Schmiedebergs Arch Pharmacol 334:412–417
22. Bönisch H (1998) Transport and drug binding kinetics in membrane vesicle preparations. In: Amara S (ed). Neurotransmitter transporters. Methods Enzymol 296:259–278
23. Eshleman AJ, Stewart E, Evenson AK, Mason JN, Blakely RD, Janowsky A, Neve KA (1997) Metabolism of catecholamines by catechol-O-methyltransferase in cells expressing recombinant catecholamine transporters. J Neurochem 69:1459–1466
24. Bönisch H, Rodrigues-Pereira E (1983) Uptake of ^{14}C-tyramine and release of extravesicular ^{3}H-noradrenaline in isolated perfused rabbit hearts. Naunyn Schmiedebergs Arch Pharmacol 323:233–244
25. Wenge B, Bönisch H (2013) The role of cysteines and histidins of the norepinephrine transporter. Neurochem Res 38:1303–1314
26. Yamamura HI, Enna SJ, Kuhar MJ (1978) Neurotransmitter receptor binding. Raven Press, New York
27. Williams LT, Lefkowitz RJ (1978) Receptor binding studies in adrenergic pharmacology. Raven Press, New York

28. Reith MEA (1997) Neurotransmitter transporters (ed.). Structure, function, and regulation. Humana, Totowa, NJ
29. Amara SG (1998) Neurotransmitter transporters (ed). Methods Enzymol 296:307-318
30. Janowsky A, Neve K, Eshleman AJ (2001) Uptake and release of neurotransmitters. Curr Protoc Neurosci Chapter 7:Unit 7.9
31. Motulsky HJ, Christopoulos A (2003) Fitting models to biological data using linear and nonlinear regression. A practical guide to curve fitting. GraphPad Sostware Inc., San Diego, CA, www.graphpad.com
32. Sitte HH, Freissmuth M (2006) Neurotransmitter transporters (eds). Handb Exp Pharmacol, vol 175. Springer, Berlin
33. Iversen LL (1965) The inhibition of noradrenaline uptake by drugs. Adv Drug Res 2:1–46
34. Iversen LL (1965) The uptake of adrenaline by the rat isolated heart. Br J Pharmacol 24:387–394
35. Burgen ASV, Iversen LL (1965) The inhibition of noradrenaline uptake by sympathomimetic amines in the rat isolated heart. Br J Pharmacol 25:34–49
36. Born GV, Gillson RE (1959) Studies on the uptake of 5-hydroxytryptamine by blood platelets. J Physiol 146(3):472–491
37. Pifl C, Giros B, Caron MG (1993) Dopamine transporter expression confers cytotoxicity to low doses of the Parkinsonism-inducing neurotoxin 1-methyl-4-phenylpyridinium. J Neurosci 13:4246–4253
38. Callingham BA (1967) The effects of imipramine and related compounds on the uptake of noradrenaline into sympathetic nerve endings. In: Garattini S, Dukes MNG (eds) Excerpta medica international congress series 122. Excerpta Medica Foundation, Amsterdam, pp 35–43
39. Maxwell RA, Ferris RM, Burcsu J, Chaplin Woodward E, Tang D, Williard K (1974) The phenyl rings of tricyclic antidepressants and related compounds as determinants of the potency of inhibition of the amine pumps in adrenergic neurons of the rabbit aorta and in rat cortical synaptosomes. J Pharmacol Exp Ther 191:418–430
40. Koe BK (1976) Molecular geometry of inhibitors of the uptake of catecholamines and serotonin in synaptosomal preparations of rat brain. J Pharmacol Exp Ther 199:649–661
41. Richelson E, Pfenning M (1984) Blockade by antidepressants and related compounds of biogenic amine uptake into rat brain synaptosomes: most antidepressants selectively block norepinephrine uptake. Eur J Pharmacol 104:277–286
42. Segel IH (1975) Enzyme kinetics. Behavior and analysis of rapid equilibrium and steady-state enzyme systems. Wiley, New York
43. Harder R, Bönisch H (1985) Effects of monovalent ions on the transport of noradrenaline across the plasma membrane of neuronal cells (PC-12 cells). J Neurochem 45:1154–1162
44. Schömig E, Korber M, Bönisch H (1988) Kinetic evidence for a common binding site for substrates and inhibitors of the neuronal noradrenaline carrier. Naunyn Schmiedebergs Arch Pharmacol 337:626–632
45. Ramamoorthy S, Ramamoorthy JD, Prasad PD, Bhat GK, Mahesh VB, Leibach FH, Ganapathy V (1995) Regulation of the human serotonin transporter by interleukin-1 beta. Biochem Biophys Res Commun 216(2):560–567
46. Kong E, Sucic S, Monje FJ, Savalli G, Diao W, Khan D, Ronovsky M, Cabatic M, Koban F, Freissmuth M, Pollak DD (2015) STAT3 controls IL6-dependent regulation of serotonin transporter function and depression-like behavior. Sci Rep 5:9009
47. Cool DR, Leibach FH, Bhalla VK, Mahesh VB, Ganapathy V (1991) Expression and cyclic AMP-dependent regulation of a high affinity serotonin transporter in the human placental choriocarcinoma cell line (JAR). J Biol Chem 266(24):15750–15757
48. Qian Y, Galli A, Ramamoorthy S, Risso S, DeFelice LJ, Blakely RD (1997) Protein kinase C activation regulates human serotonin transporters in HEK293 cells via altered cell surface expression. J Neurosci 17(1):45–57
49. Apparsundaram S, Galli A, DeFelice LJ, Hartzell HC, Blakely RD (1998) Acute regulation of norepinephrine transport: I. Protein kinase C-linked muscarinic receptors influence transport capacity and transporter density in SK-N-SH cells. J Pharmacol Exp Ther 287:733–743
50. Sucic S, Bryan-Lluka LJ (2002) The role of the conserved GXXXRXG motif in the expression and function of the human norepinephrine transporter. Brain Res Mol Brain Res 108:40–50
51. Sucic S, Packowski FA, Runkel F, Bönisch H, Bryan-Lluka LJ (2002) Functional significance of a highly conserved glutamate residue of the human noradrenaline transporter. J Neurochem 81:344–354
52. Paczkowski FA, Bryan-Lluka LJ (2001) Tyrosine residue 271 of the norepinephrine transporter is an important determinant of its pharmacology. Brain Res Mol Brain Res 97:32–42

53. Paczkowski FA, Bönisch H, Bryan-Lluka LJ (2002) Pharmacological properties of the naturally occurring Ala457Pro variant of the human norepinephrine transporter. Pharmacogenetics 12:165–173
54. Giros B, Wang Y-M, Suter S, McLeskey SB, Pifl C, Caron MG (1994) Delineation of discrete domains for substrate, cocaine, and tricyclic antidepressant interactions using chimeric dopamine-norepinephrine transporters. J Biol Chem 269:15985–15988
55. Buck KJ, Amara SG (1994) Chimeric dopamine-norepinephrine transporters delineate structural domains influencing selectivity for catecholamines and 1-methyl-4-phenylpyridinium. Proc Natl Acad Sci U S A 91:12584–12588
56. Buck KJ, Amara SG (1995) Structural domains of catecholamine transporter chimeras involved in selective inhibition by antidepressants and psychomotor stimulants. Mol Pharmacol 48: 1030–1037
57. Kitayama S, Shimada S, Xu H, Markham L, Donovan DM, Uhl GR (1992) Dopamine transporter site-directed mutations differentially alter substrate transport and cocaine binding. Proc Natl Acad Sci U S A 89:7782–7785
58. Barker EL, Blakely RD (1996) Identification of a single amino acid, phenylalanine 586, that is responsible for high affinity interactions of tricyclic antidepressants with the human serotonin transporter. Mol Pharmacol 50:957–965
59. Surratt CK, Ukairo OT, Ramanujapuram S (2005) Recognition of psychostimulants, antidepressants, and other inhibitors of synaptic neurotransmitter uptake by the plasma membrane monoamine transporters. AAPS J 7: E739–E751
60. Stöber G, Nöthen MM, Pörzgen P, Brüss M, Bönisch H, Knapp M, Beckmann H, Propping P (1996) Systematic search for variation in the human norepinephrine transporter gene: identification of five naturally occurring missense mutations and study of association with major psychiatric disorders. Am J Med Genet 67:523–532
61. Stöber G, Hebebrand J, Cichon S, Brüss M, Bönisch H, Lehmkuhl G, Poustka F, Schmidt M, Remschmidt H, Propping P, Nöthen MM (1999) Tourette syndrome and the norepinephrine transporter gene: results of a systematic mutation screening. Am J Med Genet 88:158–163
62. Shannon JR, Flattem NL, Jordan J, Jacob G, Black BK, Biaggioni I, Blakely RD, Robertson D (2000) Orthostatic intolerance and tachycardia associated with norepinephrine-transporter deficiency. N Engl J Med 342:541–549
63. Rodríguez GJ, Roman DL, White KJ, Nichols DE, Barker EL (2003) Distinct recognition of substrates by the human and Drosophila serotonin transporters. J Pharmacol Exp Ther 306(1):338–346
64. Han DD, Gu HH (2006) Comparison of the monoamine transporters from human and mouse in their sensitivities to psychostimulant drugs. BMC Pharmacol 6:6
65. Sarker S, Weissensteiner R, Steiner I, Sitte HH, Ecker GF, Freissmuth M, Sucic S (2010) The high-affinity binding site for tricyclic antidepressants resides in the outer vestibule of the serotonin transporter. Mol Pharmacol 78: 1026–1035
66. Sucic S, Dallinger S, Zdrazil B, Weissensteiner R, Jorgensen TN, Holy M, Kudlacek O, Seidel S, Cha JH, Gether U, Newman AH, Ecker GF, Freissmuth M, Sitte HH (2010) The amino terminus of monoamine transporters is a lever required for the action of amphetamines. J Biol Chem 285:10924–10938
67. Kitayama S, Mitsuhata C, Davis S, Wang J-B, Sato T, Morita K, Uhl GR, Dohi T (1998) MPP+ toxicity and plasma membrane dopamine transporter: study using cell lines expressing the wild-type and mutant rat dopamine transporters. Biochim Biophys Acta 1404:305–313
68. Guptaroy B, Fraser R, Desai A, Zhang M, Gnegy ME (2001) Site-directed mutations near transmembrane domain 1 alter conformation and function of norepinephrine and dopamine transporters. Mol Pharmacol 79(3):520–532
69. Tatsumi M, Groshan K, Blakely RD, Richelson E (1997) Pharmacological profile of antidepressants and related compounds at human monoamine transporters. Eur J Pharmacol 340:249–258
70. Lynagh T, Khamu TS, Bryan-Lluka LJ (2014) Extracellular loop 3 of the noradrenaline transporter contributes to substrate and inhibitor selectivity. Naunyn Schmiedebergs Arch Pharmacol 387:95–107
71. Scholze P, Nørregaard L, Singer EA, Freissmuth M, Gether U, Sitte HH (2002) The role of zinc ions in reverse transport mediated by monoamine transporters. J Biol Chem 277(24):21505–21513
72. Karlin A, Akabas MH (1998) Substituted-cysteine accessibility method. Methods Enzymol 293:123–145
73. Javitch JA (1998) Probing structure of neurotransmitter transporters by substituted cysteine accessibility methods. Methods Enzymol 296:331–346
74. Ferrer JV, Javitch JA (1998) Cocaine alters the accessibility of endogenous cysteines in puta-

tive extracellular and intracellular loops of the human dopamine transporter. Proc Natl Acad Sci U S A 95:9238–9243

75. Chen J-G, Rudnick G (2000) Permeation and gating residues in serotonin transporter. Proc Natl Acad Sci U S A 97:1044–1049

76. Sucic S, Bryan-Lluka LJ (2005) Roles of transmembrane domain 2 and the first intracellular loop in human noradrenaline transporter function: pharmacological and SCAM analysis. J Neurochem 94(6): 1620–1630

Chapter 2

Tracer Flux Measurements to Study Outward Transport by Monoamine Neurotransmitter Transporters

Thomas Steinkellner, Felix P. Mayer, Tina Hofmaier, Marion Holy, Therese Montgomery, Birgit Eisenrauch, Michael Freissmuth, and Harald H. Sitte

Abstract

The physiological role of neurotransmitter transporter (NTT) proteins is the reuptake of released neurotransmitter from the synaptic cleft. NTTs accomplish uptake by undergoing a transport cycle, which relies on a return step in the empty state. In addition, NTTs can also run in the reverse direction and transport substrates out of the cells. This can be observed under conditions, where the transmembrane sodium gradient dissipates, e.g., if sodium accumulates within the cell. This reverse transport mode is also induced by amphetamines and the exact mechanism underlying the amphetamine action is still enigmatic and involves complex regulatory processes. In the current chapter, we describe various methods that can be used to assess the efflux of neurotransmitter from cells heterologously expressing the NTTs of interest or from preparations derived from intact brain tissue.

Key words Carrier-mediated efflux, Transport reversal, Neurotransmitter transporter, Superfusion, Radiolabeled tracer flux, Heterologous cell expression systems, Synaptosomes, Brain slices

1 Introduction

Three different possibilities exist to terminate synaptic transmission: (1) diffusion of neurotransmitter out of the synaptic cleft, (2) enzymatic degradation, or (3) reuptake by neurotransmitter transporters (NTT; [1]). The latter results in reaccumulation of neurotransmitters; the sodium gradient provides the driving force for reuptake from the synaptic cleft. Thus, it is by definition a secondary-active transport [2], which affords an economical and rapid reuse of released neurotransmitter. The monoamine transporters of the solute carrier 6 family (SLC6) comprise the transporters for dopamine, DAT (SLC6A3), norepinephrine, NET (SLC6A2), and serotonin, SERT (SLC6A4) [3], also abbreviated as 5-HTT [4]. The monoamine NTT family is of clinical

Heinz Bönisch and Harald H. Sitte (eds.), *Neurotransmitter Transporters: Investigative Methods*, Neuromethods, vol. 118, DOI 10.1007/978-1-4939-3765-3_2, © Springer Science+Business Media New York 2016

importance since they serve as target to alleviate or effectively treat a number of psychiatric disorders including depression and attention deficit hyperactivity disorder [5].

Chemical neurotransmission was established by the pioneering experiments of Otto Loewi: electrical stimulation of a nerve released a diffusible neurotransmitter—in Loewi's case the "Vagusstoff," i.e., acetylcholine, and this principle was shown to be universally true in both the peripheral and the central nervous system. It took some 40 years until Axelrod and Hertting documented that neurotransmission by monoamines (and most other released neurotransmitters) was terminated by a transport process rather than enzymatic degradation. [6]. Several models were proposed to account for the ability of a protein to translocate a solute over the lipid bilayer: Jardetzky condensed these ideas into a concept, which posited an alternating access mechanism [7]. This model was vindicated by the X-ray crystal structures of many transporters, including those of the bacterial transporter LeuTAa, which revealed several conformational states consistent with the sequence of events postulated by the alternating access model [8–10]. The fact that LeuTAa is highly homologous to NTTs allows for educated guesses on the mechanistic details of the transport process [11–14] and provides a reference framework for dynamic studies at the single-molecule level [15, 16].

Importantly, reuptake from the extracellular space to the cytosol is the major but not the only possible transport direction: changes in the intracellular milieu, in particular the sodium concentration, can result in reverse transport and thus lead to NTT-mediated efflux. Reverse transport can also be observed after the administration of sympathomimetic amines such as tyramine or amphetamine-like drugs [6, 17–21]. In vivo, reverse transport can be detected by microdialysis in the awake animals [22–24] and by high-speed chronoamperometry [25]. As an alternative approach, reverse transport by psychoactive amines has also been extensively studied in brain slices or synaptosomes [26–30]. However, the interpretation of mechanistic studies is confounded by two inherent limitations: (1) both slices and synaptosomes often contain several different NTTs (e.g., all monoamine transporters are included in a striatal slice preparation); (2) in addition, slices—and to a lesser extent synaptosomes—also contain the complete machinery for synaptic vesicle exocytosis [31, 32]. Accordingly, isolation of a single transporter requires the others to be blocked by specific NTT-blockers (to detect only effects mediated by the transporter of interest) and/or receptor inhibitors (because autoreceptors may be stimulated by the psychoactive compounds under study and regulate both vesicular and carrier-mediated release).

The cloning of the monoamine NTT cDNA's made it possible to heterologously express a specific monoamine transporter in appropriate cell lines such as human embryonic kidney 293 cells (HEK293). Thereby, the problems are eliminated, which arise from interferences with vesicular storage and the stimulation of monoaminergic receptors: these paradigms have been used to study reverse transport induced by any substrates of NSS members (e.g., dopamine, tyramine, amphetamine and its derivatives or others [33–37]).

In this chapter, we outline the techniques, which have been successfully used to assess reverse transport mediated by monoamine transporters. Initially, we provide the methodologies to prepare brain slices and synaptosomes and subsequently describe the procedures to work with heterologous expression systems in both static and dynamic systems, i.e., using batch release and a superfusion system, respectively. After a description of the superfusion system, we then elaborate on the experimental procedures of the release experiment. The next part focuses on the evaluation and interpretation of the data. Finally, we will discuss troubleshooting issues and point out limitations and drawbacks of the methods.

2 Materials

2.1 Synaptosome Preparation from Animal Brain Tissue

Krebs-HEPES buffer or Krebs–Henseleit buffer:

Krebs-HEPES buffer (KHB): 25 mM HEPES, 120 mM NaCl, 5 mM KCl, 1.2 mM $CaCl_2$, and 1.2 mM $MgSO_4$ supplemented with 5 mM d-glucose, adjusted with NaOH to pH = 7.4.

Krebs–Henseleit buffer (KHensB): 118 mM NaCl, 4.7 mM KCl, 1.2 mM $MgSO_4$, 1.25 mM $CaCl_2$, 1.2 mM KH_2PO_4, 25 mM $NaHCO_3$, 11 mM glucose.

Prepare KHensB freshly every day and oxygenate with 95% O_2/5% CO_2 for 1 h before use to adjust pH to 7.4.

Animal brain removed from either mouse or rat.

Tissue douncer with appropriately sized teflon-coated pestle for the preparation of synaptosomes.

Phosphate-buffered saline (PBS) containing protease inhibitors (Roche Complete™).

Protein determination kit (e.g., BCA kit, Pierce/Thermo Scientific).

24-well plates.

Whatman GF/B filters (1 mm diameter).

2.2 Slice Preparation from Animal Brain Tissue

Krebs-HEPES buffer or KHensB buffer (see above for composition).

Animal brain removed from either mouse or rat.

McIlwain tissue chopper (Fig. 1) for the preparation of brain slices.

2.3 HEK293 Cell Culture and Reagents

DMEM medium (any source is good provided that they adhere to the original recipe, e.g., Gibco, Invitrogen Life Science, Bethesda, MD).

Fetal bovine serum (FBS).

Fig. 1 A McIlwain tissue chopper, ready for use

0.05 % Trypsin/EDTA (Sigma-Aldrich, St. Louis, MO).

l-Glutamine (Gibco, Invitrogen Life Science, Bethesda, MD).

10 cm or 15 cm dishes; 24-well or 96-well culture plates (Greiner, Sarstedt, BD Biosciences, Falcon).

Penicillin (10,000 U/mL) and streptomycin (10 mg/mL) solutions are frozen at −20 °C (Gibco, Invitrogen Life Science); 5 mL is added to 0.5 L of DMEM complete culture medium.

Cell line: HEK293 cells transiently or stably expressing NTT of interest.

0.1 mg/mL poly-d-lysine solution in sterile H_2O.

Round coverslip (5 mm diameter).

2.4 Equipment, Software, and Accessories

Cell culture incubator, 37 °C, 5 % CO_2.

Sterile hood for the handling of the cell lines.

Vacuum pump for cell washes.

A personal computer with appropriate data calculation and graphics software.

For the identification of the brain structures, use pertinent brain atlases (e.g., Franklin and Paxinos for mouse brain or the rat brain atlas by König and Klippel). Note: for the preparation of synaptosomes and brain slices, brains need preferentially to be dissected freshly and on a cold plate (not frozen).

3 Methods

In the methods section, we will describe in the first paragraphs how the three preparations covered in this chapter are produced (Sects. 3.1–3.3) before we describe the superfusion assay (Sect. 3.4) and the evaluation of the data (Sect. 3.5).

3.1 Preparation of Rodent Brain Synaptosomes

Kill mouse or rat by cervical dislocation or decapitation.

Remove the brain carefully (Note: make sure that the meninges are removed thoroughly otherwise you risk serious damage to the delicate brain tissue during dissection) and keep on ice.

Dissect both left and right region(s) of interest.

Homogenize in ice-cold 0.32 M sucrose in phosphate-buffered saline (PBS) containing protease inhibitors (Roche Complete™).

Centrifuge the suspension for 10 min at 1000 × *g*. Keep the supernatant, discard the pellet.

Centrifuge the supernatant for 15 min at 12,600 × *g*. Discard the supernatant and resuspend the pellet (referred to as P_2) in KHB.

Measure the wet weight of P_2 or determine the protein concentration of P_2 after resuspension in buffer to estimate the amount of synaptosomes.

Use the resulting synaptosomes directly or freeze at −80 °C. Synaptosomes can be stored at −80 °C for at least 2 years with only minor loss in functional activity.

3.2 Preparation of Mouse Brain Slices

Retrieve a rodent brain as described above and place it into ice-cold buffer (KHB or KHensB).

Dissect the regions of interest (e.g., striatum, hippocampus, or cortical tissue) and store in ice-cold buffer in a watch glass on ice.

Take dissected brain region and place onto a Whatman filter paper saturated with ice-cold buffer; place filter with brain onto the cutting stage of the McIlwain tissue chopper and operate the knife slowly to cut the first tissue slice (usually, 0.3 mm thickness is recommended). Discard the first section and start collecting subsequent sections using a fine brush by gently swiping them and placing them in ice-cold buffer for storage (Fig. 2).

Repeat until all slices of your region of interest are retrieved.

Handle the slices cautiously with the brush avoiding damage to the tissue.

Tissue slices can be kept in ice-cold buffer for a couple of hours but tissue quality and integrity decrease with time.

3.3 Heterologously Expressing Cell Lines

The expression of the NTT of interest in appropriate cell lines can be bothersome from time to time and expression levels may vary significantly. This may cause differences in results for NTT-mediated efflux from lab to lab but even among different cell lines

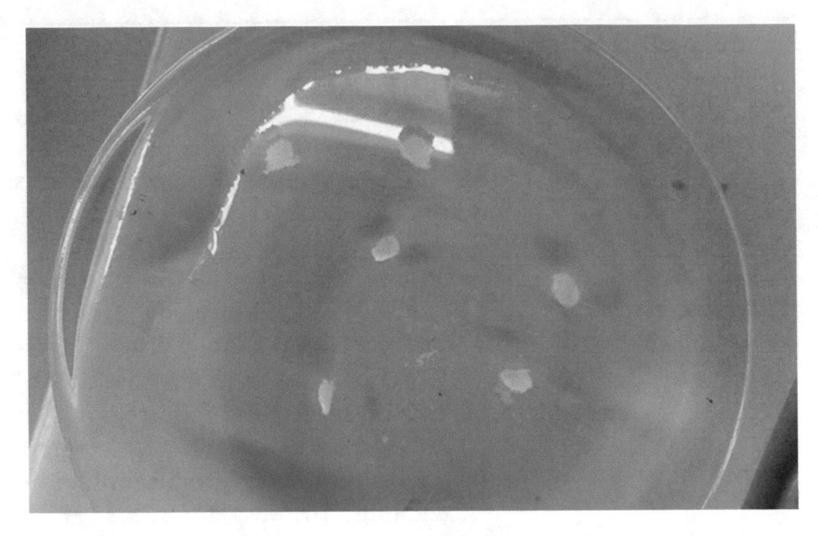

Fig. 2 Watch glass holding ice-cold buffer with floating striatal sections prepared from rat brain tissue

within one lab. The reason for the differences might simply be that the properties of the cell lines differ considerably with regard to the expression of kinases and other important regulatory factors, e.g., protein kinase C, [38], or αCamKII, [39–43]. Hence, it would be good to start the approach by screening the literature and selecting the right cell line based on the experience of other groups. For instance, DAT has been expressed in a number of different cell lines, including HEK293 cells [32], PC-12 cells, [44], SK-NMC cells [45] and LLC-PKC1 cells [46, 47]. All cell lines have their unique advantages: HEK293 cells for instance can be most easily transfected and afford high expression levels. In contrast, LLC-PKC1 cells support much lower expression levels but these cells are endowed with a large complement of protein kinases. Clearly, it depends on the goal, which is being pursued: the high expression in HEK293 cells is best suited to study transporter-ligand interactions while the low expression profile of LLC-PKC1 cells can be utilized to assess the influence of kinases on transporter regulation [47].

For transfection of HEK293 cells we usually resort to the $CaPO_4$ transfection method since it is cheap, easy and reliable; only if this method does not lead to the desired cell expression result, we use a lipofection method. In the following, we outline how we perform $CaPO_4$ transfections, followed by one lipofection method which is given as an example for a large variety of different commercially available products. Still, the number of products is increasing and it provides a simple and efficient way of transfection. However, lipofection interferes with the lipidome of the cells and may therefore interfere with lipid-transporter interactions [48, 49].

At the very least, these aspects must be considered, when selecting the transfection method.

Ca2+ Transfections

HeBS buffer (2×):	
HEPES	50 mM
NaCl	280 mM
$Na_2HPO_4 \cdot 2H_2O$	1.5 mM

Adjust the pH to 7.08 with 10 N NaOH; accurate pH is critical for efficient transfection. Sterilize by filtration through a 0.45-μm nitrocellulose filter. Store at −20 °C. Thaw only once!

Phosphate buffered saline (PBS):	
KCl	2.7 mM
KH_2PO_4	1.5 mM
NaCl	137 mM
$Na_2HPO_4 \cdot 2H_2O$	4.3 mM

Dissolve in sterile water (Milli-Q), adjust to pH 7.3–7.4 (by adding 10 N NaOH).

Seed cells to achieve approximately 40% confluence in a 10 cm dish on the day of transfection. Mix reagents as follows ($CaCl_2 + H_2O = 1:8.6$):

H_2O	430 μL	
$CaCl_2$	(2.3 M)	50 μL
DNA (1 μg/μL)	20 μL	
HeBS (2×)	500 μL	

Incubate the mixture for 6 min at room temperature to allow the reaction to take place: here, DNA-Ca^{2+} should form a fine precipitate. The precipitate should not be too coarse. Drop the suspension onto the medium covering the cells.

Apply the "glycerol shock" after 4–6 h of incubation time at 37 °C to increase transfection efficiency: remove the transfection media, add 1 mL of glycerol shock solution (thoroughly mix 13.8 mL glycerol and 86.2 mL PBS, sterile filtered afterwards) and remove it immediately after and rapidly by aspiration.

Wash the cells with 10 mL PBS and supply 10 mL of fresh, warm media.

Wait for at least 24 h before performing experiments on transiently expressing cells. To establish a stably expressing cell line, start the selection process with the appropriate antibiotic (e.g., geneticin) 48–72 h after the glycerol shock.

Turbofection

Seed cells into 10 cm dishes and use them for turbofection when they are approximately 80–90 % confluent.

Combine 1 μg of DNA of interest with "empty" vectors (e.g., pcDNA3.1) to reach a total amount of 5 μg of DNA (Note: dilution of transporter DNA is usually recommended as lipofection can induce massive overexpression of the protein of interest; in order to achieve more "physiological" expression, dilution series of transporter DNA should be performed and tested).

Add 500 μL of DMEM to the DNA mix followed by the addition of 5 μL Turbofect (Fermentas; vortex Turbofect well before adding).

Vortex the mixture and incubate for 15–20 min at room temperature.

In the meantime, remove the media from the cells and replace with 4.5 mL of fresh media including FCS and antibiotics.

After the incubation of Turbofect–DNA mixture is ready, add the solution to the cells. Incubate the cells for another 24–48 h at 37 °C.

It is advisable to first verify that the monoamine NTT of interest is expressed in the given cell line at adequate levels prior to carrying out a release assay. Expression can be confirmed by performing an uptake assay or by labeling the NTT of interest with a fluorescent protein (e.g., GFP or mCherry). Note: the addition of a fluorescent protein can compromise NTT function; for instance, only SERT tolerates the addition of a fluorescent protein to the carboxyl terminus [38], whereas in all other monoamine NTT surface expression is substantially reduced because their C-terminal PDZ-binding motif cannot be masked [50, 51].

Intuitively, stable transfection is more appealing, but it may not be necessary to have a cell line with a homogeneous expression level. In addition, it may not be possible to express a transporter of interest in a stable manner. This is for instance true for mutants, which generate a large leak current. A point in case is the mutant DAT-Y335A [52]; this mutation converts the transporter into a ion-channel-like pore [53]. Here, stable expression has not been possible, because the large leak conductance of DAT-Y335A apparently precluded long term cell survival (unpublished results).

3.4 Release Assays

While the preparations (synaptosomes, brain slices or cells on coverslips or in plates) differ, the superfusion is done in a very similar manner; the variations are modest. The static batch release assay will be described in Sect. 3.4.2.

3.4.1 Superfusion Assay

We will shortly describe the superfusion apparatus and how to set it up before each experiment before the actual experiments involving the three different preparations will be explained.

The system is designed to allow for rapid removal of the released neurotransmitter by continuous superfusion of the preparation of interest. The goal is to preclude reuptake of the released neurotransmitter (regardless of whether released spontaneously or after a stimulus) by the cognate NTT or other transporters. An initial superfusion is employed to define the baseline, i.e., to estimate the spontaneous release or leak in the respective preparation.

Our superfusion apparatus was designed to comprise 12 individual channels (Fig. 3). Hence, 12 brain slices, 12 synaptosomal fractions or 12 cell coverslips can be used per assay in parallel. This allows for an adequate number of replicates; we rely on triplicate determinations for each experimental condition. A schematic representation of a superfusion chamber is given in Fig. 4b. Of importance to control the temperature in the chambers, the tubing ought to be immersed in a water bath, which should be kept at a temperature higher than the desired value in the chamber to yield a temperature of 25 °C in the chambers proper.

The preparations are continuously superfused with buffer at a flow rate of 0.7 mL/min. It is essential that no air bubbles are trapped in the superfusion tubings. Air bubbles can be easily removed by rinsing the superfusion system with 30 % isopropanol in ddH_2O before the experiment. After the isopropanol, the system is washed extensively with ddH_2O (10 min) to remove any residual isopropanol before equilibrating the system buffer (10 min).

The superfusate must obviously have a constant temperature to eliminate a source of variability. The control of the temperature is achieved by immersing the tubing (total diameter = 1 mm, a luminal diameter = 0.35 mm) over a length of 60 cm in a water bath set at the pertinent temperature (usually 25 °C). Note that

Fig. 3 Twelve-channel superfusion apparatus. Note that the valves in front allow for rapid and simple exchange of solutions, i.e., from baseline buffers to drug containing buffers. The tubing is typically immersed in a water bath, which should be kept at a temperature higher than the desired value in the chamber

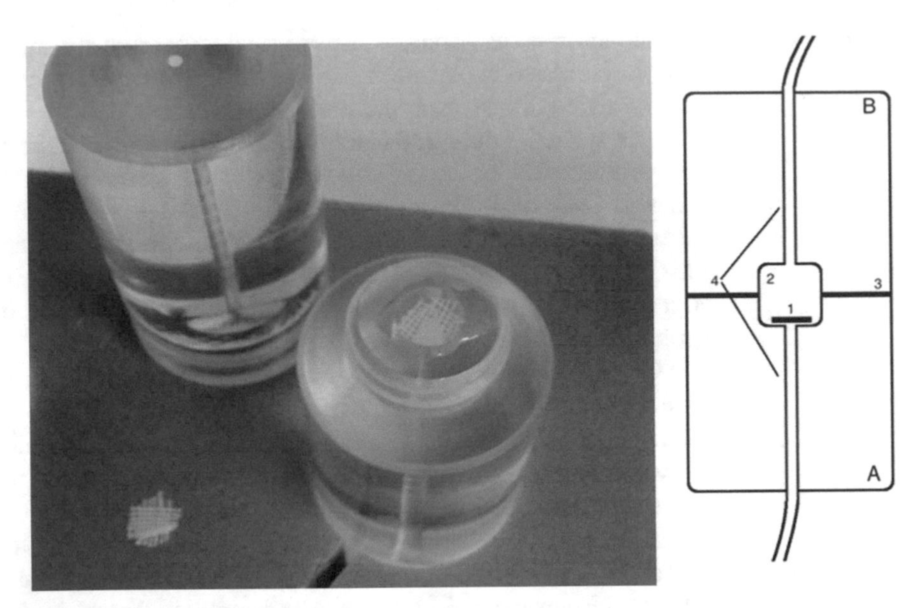

Fig. 4 (**a**) Detailed view into the superfusion chamber. The volume of the central chambers is 200 μL and contains a grid which holds the slices, glass coverslips or synaptosome-loaded filters. (**b**) Schematic representation of the superfusion system. The investigated material (*1*) is placed into the superfusion chamber (*2*; volume = 200 μL). A rubber gasket (*3*) ensures tight closure of the two hemispheres (A and B with a total dimension of 8 cm × 2.5 cm). The flow direction of the superfusion buffer is from A to B via channels (*4*) cut into the hemispheres. Adapted from Singer [68]

the flow rate of the buffer results in mechanical stress for the tested preparations. The flow-related shear forces are limited by the size of the chambers, which harbor the preparations: their volume is 200 μL with a diameter of 8 mm.

For superfusion of synaptosomes, synaptosomal P_2 pellets are resuspended in the respective buffer to achieve a concentration of 1 mg wet weight/15 μL or 100 μg of protein.

For superfusion of brain slices, individual sections are used per channel.

Synaptosomes are incubated with a [^{3}H]-labeled substrate of interest: for instance when DAT is the NTT in focus, [^{3}H]MPP+ or [^{3}H]dopamine can be used to load the tissue. When SERT is the NTT of interest, [^{3}H]5-HT should be used.

CAVEAT: biogenic amines are substrates for monoamine oxidases and prone to spontaneous oxidation. Hence, they can get rapidly metabolized in native tissue such as synaptosomes or tend to spontaneously oxidize. It is therefore advisable to block MAOs throughout the assay by adding for instance 100 nM pargyline (a MAO blocker) and to prevent spontaneous oxidation by including 100 nM ascorbate (as antioxidant) in the buffer.

CAVEAT: radiolabeled substrates may be taken up by multiple transporters, e.g., [^{3}H]dopamine is not only a substrate for DAT but also for NET and SERT. Hence, when DAT is the transporter of interest, NET and SERT should be blocked. We therefore add 100 nM desipramine (NET blocker) and 100 nM paroxetine to the buffer. In contrast, when SERT is the transporter of interest, we add 100 nM nomifensine to the buffer, which blocks both DAT and NET at similar potencies. If NET is to be investigated, the addition of a 100 nM GBR12909 or GBR12935 (specific DAT inhibitors) and 100 nM paroxetine is recommended.

Preincubation time for synaptosomes and slices:

Preincubate synaptosomes and slices for 30 min at 37 °C with [^{3}H]-labeled substrate (e.g., 100 nM [^{3}H]dopamine or 100 nM [^{3}H]5-HT) in working buffer (+100 nM pargyline+100 nM ascorbate+respective blockers for other NTT, e.g., 100 nM desipramine (to block NET) and 100 nM paroxetine (to block SERT) or 100 nM nomifensine to block DAT and NET, respectively).

Preincubation time for cells:

Preincubate cells for 20 min at 37 °C with [^{3}H]-labeled substrate in buffer (depending on the cell line using, addition of pargyline to block MAO or ascorbate as an antioxidant is recommended. Also, if neuronal-like cell lines are used, it might be advisable to add the respective inhibitors for NTT that are not the primary focus of investigation, e.g., both SHY5Y cells and PC12 cells endogenously express NET)

Insert preparations into superfusion apparatus and perform washout:

After preincubation, preparations are transferred to the superfusion apparatus.

To insert preparations, stop perfusion flow and open all the channels.

Synaptosomes: prepare a plastic dish with GF/B filters (8 mm diameter) and pipet 15 μL of preincubation solution onto Whatman GF/B filters. Immediately put one loaded filter into one channel, close the channels and perfuse with buffer.

Slices: insert one slice per channel.

Cells: insert one coverslip per channel.

After the preparations have been placed into the chambers, a washout phase is started to equilibrate the system and to establish a stable baseline efflux of radioactivity. Usually a washout time of 45 min is sufficient (keep in mind that MAO blocker, ascorbate and inhibitors of the other NTTs should also be present in the buffer during this washout phase).

During washout, prepare the required amount of scintillation vials in a rack and fill with 2 mL scintillation cocktail (we usually use 50 mL scintillation vials filled with 2 mL of scintillation liquid).

Also during the washout period, prepare the working solutions needed for the efflux experiment, i.e., prepare the releaser solution:

For example "*releaser*" *solution* (for DAT, we usually use 3–10 μM of d-amphetamine, for SERT 3–10 μM of *para*-chloroamphetamine [*p*CA] in working buffer since this allows for maximal substrate release at monoamine transporters; keep in mind that amphetamines usually have a bell-shaped dose-response curve, i.e., further increases in d-amphetamine or *p*CA concentrations do not increase efflux but rather have inhibitory effects on release; see Seidel et al. [38] for details).

Note: depending on the transporter under investigation, one should choose the respective amphetamine: amphetamines usually are promiscuous molecules but they still show preferences for the one or other transporter. For instance, d-amphetamine has higher affinities for DAT and NET, whereas *p*CA shows an increased affinity for SERT. Hence, it is advisable to select the most potent amphetamine for a given transporter under investigation, because this minimizes off-target effects.

"*Releaser + inhibitor*" (*control*): another way to test, whether release through the NTT of interest is specific it is recommended to include a condition where release is blocked by the pertinent inhibitor for this transporter, e.g., 3–10 μM d-amphetamine + 1 μM GBR12909 for DAT.

After washout, the actual experiment is started by collecting at least 3 × 2-min fractions where the preparation is superfused with buffer only. After the first three fractions, the tubings can be switched to the releaser solution. When also looking at the effect of an inhibitor, we usually collect three 2-min with inhibitor only (e.g., 1 μM GBR12909) or buffer only before collecting five 2-min fractions with releaser and releaser + inhibitor, respectively.

At the end of the experiment, synaptosomes or slices are collected in tubes filled with 2 mL of 1 % SDS to recover the remaining amount of [^{3}H] substrate present in the tissue.

Superfusion experiments with cells grown on coverslips are terminated by collecting three 2-min fractions where the superfusion system is washed with 1 % SDS to lyse the cells.

At the end of the experiment, put the lid onto the scintillation vials and shake tubes well before measuring in the liquid scintillation counter.

3.5 Data Analysis

Calculations are based as fractional release: hence the first step is to calculate the radioactivity initially present in the preparation, which is sum of the radioactivity present in all collected 2-min fractions and radioactivity present at the end, i.e., the radioactivity released after solubilization of the tissue with 1 % SDS. The released radioactive [^{3}H]substrate in any given 2-min fraction is expressed as percentage of the radioactivity present in the slice at the start of this very collection period, i.e., fractional release = $cpm_{2\text{-min fraction}}$/(total radioactivity – sum of previously release) × 100.

Usually the baseline efflux of radioactivity is stable and amounts to about 1–1.5 % fractional release per 2-min buffer superfusion. Efflux rises to 5–10 % after the addition of amphetamine (Fig. 5).

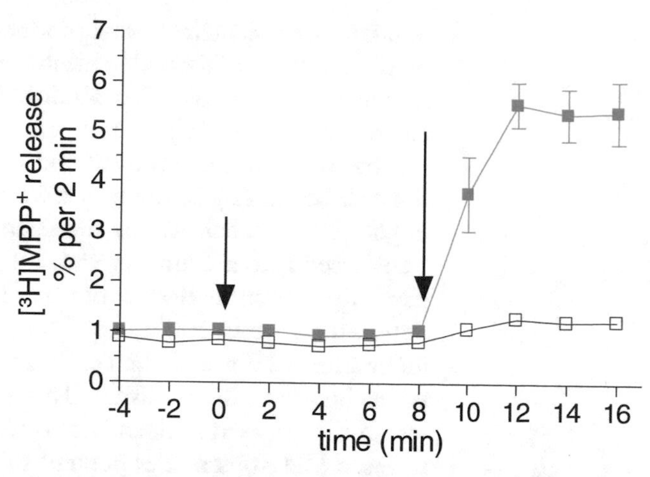

Fig. 5 Superfusion assay, evaluated and displayed: representative experiment. Cells expressing human DAT were preloaded with [^{3}H]MPP^{+} and superfused until a stable baseline was reached. The experiment was started with the collection of 2-min fractions. After three fractions (min-6 until 0 min; *first arrow*) of basal efflux, cells were exposed to cocaine (10 μM), or left at control conditions as indicated. After four fractions (from 8 min onwards; *second arrow*), amphetamine (10 μM) was added to all superfusion channels. After four fractions (i.e., after 16 min), all channels were switched to SDS conditions and the remaining radioactivity lysed from the cells. Data are presented as fractional efflux, i.e., each fraction is expressed as the percentage of radioactivity present in the cells at the beginning of that fraction. Symbols represent means ± S.E.M. of three observations (one observation equals one superfusion chamber)

3.5.1 Static Batch Release Assay

The problem of the batch release assay format is its static nature: Here, the diffusion of substrate and of the compounds under investigation cannot be controlled like in the superfusion apparatus. Superfusion provides a robust assay format to assess transporter-mediated efflux, because confounding effects arising form back diffusion are eliminated [54]. The disadvantage of superfusion, however, is the large volume of superfusate and hence the need of a larger amount of the compound under study. Therefore, static batch release is the preferred method when the amount of compound under scrutiny is small [55], but its limitations must be kept in mind. In the following, we will outline how the static release assay has worked out well in our hands.

Cells are grown in poly-d-lysine-coated 96-well plates (4×10^4 cells per well). The cells are preloaded with 0.05 μM [^{3}H]substrate for 20 min at 37 °C in a final volume of 0.1 mL/well. The residual extracellular radioactivity is removed by three gentle wash steps with Krebs-Ringer-HEPES buffer. The cells are then incubated with the test compounds at room temperature and compared to a reference compound (for instance, d-amphetamine can be used for all three monoamine transporters). All compounds are used at the concentration, which results in a 50% inhibition of substrate uptake. The specificity of drug-induced release needs to be assessed by the addition of a specific inhibitor at appropriate concentration (for instance, 10 μM of mazindole for DAT and NET or 10 μM paroxetine for SERT) to the test compound. After 10 min, the incubation buffer is removed and transferred into a counting vial; the cells remaining in the well are overlaid with a solution containing 1% SDS to extract the radioactivity. The resulting solution is transferred into a counting vial. All samples are subjected to standard liquid scintillation counting. It is preferable to perform all determinations in triplicates. The sum of the radioactivity in the incubation buffer and the cell lysate represents the total [^{3}H]substrate included in the assay. This sum is the reference value, to which the released radioactivity is related: the data are expressed as released [^{3}H]substrate as percent of total available radioactivity.

4 Notes

The application of the above mentioned superfusion apparatus provides a robust technique to measure NTT-mediated reverse transport. However, various circumstances must be taken into consideration, which might result in flawed data interpretations.

1. Liquid scintillation counting is our method of choice to determine the amount of tritium within the superfusates. Liquid scintillation cocktails (e.g., Rotiscint®) convert the energy of the beta-decay of tritium into light signals. Some drugs may absorb photons emitted by scintillators and quench the cognate signal. Therefore, one has to ensure that the applied substances do not influence the readout at the tested concentrations. For example, a substance-related quench would seemingly reduce the amount of released tritiated substrate and could be misinterpreted as transporter-related effect.
2. Uptake inhibition assays do not provide the possibility to differentiate between non-transported competitive NTT inhibitors or amphetamine-like releasers [56]. In general, amphetamines are accepted as NTT substrates and compete with the endogenous substrate for the binding site. As a result, amphetamines dose-dependently inhibit uptake of tritiated substrates. The use of the superfusion technique allows bypassing this limitation of inwardly directed radiotracer flux assays. The presence of amphetamines enhances the efflux of tritiated substrates whereas non-transported inhibitors, e.g., cocaine, do not result in NTT-mediated reverse transport [57]. In addition, NTTs utilize the preexisting sodium gradient as driving force for their concentrative transport.

Disrupting the sodium gradient by addition of the Na^+/H^+ ionophore monensin [58] will selectively enhance efflux triggered by true substrates/amphetamines [56, 59]. However, the use of monensin alters the intracellular pH, as monensin serves as a leak for intracellular protons. Serotonin (5-HT) is weak base with a p*K*a value of 10.4 [60]. A reduction in intracellular H^+ inevitably increases the deprotonated fraction of 5-HT, which can passively diffuse across the plasma membrane. This results in a transporter-independent "pseudo efflux" [59]. It is worth mentioning that inhibition of NTTs by various inhibitors can unmask a basal loss of substrates by diffusion. Biogenic amines do have a limited—albeit measurable—potential to cross the plasma membrane by passive diffusion [61]. Therefore, if results obtained from superfusion experiments with [^{3}H]-5HT or [^{3}H]-DA are ambiguous, [^{3}H]-MPP^+ should be used as NTT substrate. MPP^+ carries a permanent charge and passive diffusion is thus negligible [54]. The cells or the synaptosomal preparation under investigation may also express additional gradient-driven transporters: Members of the SLC22 family, epitomized by organic cation transporters 1–3 (OCTs, SLC22A1–3) or the plasmalemmal monoamine transporter (PMAT, SLC29A4), translocate DA, NE, 5-HT, and MPP^+ across cellular membranes [62, 63]. Their coexistence with SLC6 family members may confound the interpretation of outwardly directed transport of tritiated biogenic amines or MPP^+. OCTs and PMAT are very broadly expressed; they can be found in both peripheral tissue and in the central nervous system, where they are expressed in neuronal and non-neuronal cell types [63–67]. As a consequence, gradient-driven transporters may contribute to efflux measured during superfusion experiments. OCTs and PMAT are not blocked by most NTT-inhibitors. Therefore, in the presence of OCTs and/or PMAT, inhibition of NTTs can enhance a basal loss of substrate, because the compensatory re-uptake by NTTs is eliminated by the NTT-inhibitor. This effect is pronounced in systems containing vesicular storage pools and/or upon application of amphetamine-like releasers. Efflux triggered by substances targeting NTTs, in particular their blockers, should be interpreted cautiously, if there is evidence for the presence of other gradient-driven transporters. As a safeguard, we recommend the use of the OCT and PMAT blocker decynium-22 (D22). D22 exhibits rather low affinities for NTTs. Therefore, as control experiment, the experimenter might test, if the efflux in presence of NTT blockers is sensitive to increasing concentrations of D22.

Acknowledgements

The authors wish to thank the Austrian Science Fund for continuous support (grant F35).

References

1. Iversen LL (1971) Role of transmitter uptake mechanisms in synaptic neurotransmission. Br J Pharmacol 41:571–591
2. Rudnick G, Clark J (1993) From synapse to vesicle: the reuptake and storage of biogenic amine neurotransmitters. Biochim Biophys Acta 1144:249–263
3. Kristensen AS, Andersen J, Jorgensen TN et al (2011) SLC6 neurotransmitter transporters: structure, function, and regulation. Pharmacol Rev 63:585–640
4. Nelson N (1998) The family of Na+/Cl– neurotransmitter transporters. J Neurochem 71:1785–1803
5. Iversen L (2000) Neurotransmitter transporters: fruitful targets for CNS drug discovery. Mol Psychiatry 5:357–362
6. Axelrod J, Whitby LG, Hertting G (1961) Effect of psychotropic drugs on the uptake of 3 H-Norepinephrine by tissues. Science 133:383–384
7. Jardetzky O (1966) Simple allosteric model for membrane pumps. Nature 211:969–970
8. Singh SK, Piscitelli CL, Yamashita A et al (2008) A competitive inhibitor traps LeuT in an open-to-out conformation. Science 322:1655–1661
9. Singh SK, Yamashita A, Gouaux E (2007) Antidepressant binding site in a bacterial homologue of neurotransmitter transporters. Nature 448:952–956
10. Yamashita A, Singh SK, Kawate T et al (2005) Crystal structure of a bacterial homologue of Na+/Cl−-dependent neurotransmitter transporters. Nature 437:215–223
11. Forrest LR, Rudnick G (2009) The rocking bundle: a mechanism for ion-coupled solute flux by symmetrical transporters. Physiology (Bethesda) 24:377–386
12. Forrest LR, Zhang YW, Jacobs MT et al (2008) Mechanism for alternating access in neurotransmitter transporters. Proc Natl Acad Sci U S A 105:10338–10343
13. Penmatsa A, Gouaux E (2014) How LeuT shapes our understanding of the mechanisms of sodium-coupled neurotransmitter transporters. J Physiol 592:863–869
14. Shi L, Quick M, Zhao Y et al (2008) The mechanism of a neurotransmitter:sodium symporter—inward release of Na+ and substrate is triggered by substrate in a second binding site. Mol Cell 30:667–677
15. Zhao Y, Terry D, Shi L et al (2010) Single-molecule dynamics of gating in a neurotransmitter transporter homologue. Nature 465:188–193
16. Zhao Y, Terry DS, Shi L et al (2011) Substrate-modulated gating dynamics in a Na+-coupled neurotransmitter transporter homologue. Nature 474:109–113
17. Barger G, Dale HH (1910) Chemical structure and sympathomimetic action of amines. J Physiol 41:19–59
18. Tainter ML, Chang DK (1927) The antagonism of sympathetic and adrenaline content of the spleen, kidney, and salivary glands in the sheep. J Pharmacol Exp Ther 30:193–207
19. Furchgott RF, Kirpekar SM, Rieker M et al (1963) Actions and interactions of norepinephrine, tyramine and cocaine on aortic strips of rabbit and left atria of guinea pig and cat. J Pharmacol Exp Ther 142:39–58
20. Ross SB, Kelder D (1977) Efflux of 5-hydroxytryptamine from synaptosomes of rat cerebral cortex. Acta Physiol Scand 99:27–36
21. Glowinski J, Axelrod J (1965) Effect of drugs on the uptake, release, and metabolism of H3-norepinephrine in the rat brain. J Pharmacol Exp Ther 149:43–49
22. Agneter E, Sitte HH, Stockl-Hiesleitner S et al (1995) Sustained dopamine release induced by secretoneurin in the striatum of the rat: a microdialysis study. J Neurochem 65:622–625
23. Gainetdinov RR, Fumagalli F, Jones SR et al (1997) Dopamine transporter is required for in vivo MPTP neurotoxicity: evidence from mice lacking the transporter. J Neurochem 69:1322–1325
24. Gainetdinov RR, Jones SR, Fumagalli F et al (1998) Re-evaluation of the role of the dopamine transporter in dopamine system homeostasis. Brain Res Brain Res Rev 26:148–153
25. Daws LC, Toney GM, Davis DJ et al (1997) In vivo chronoamperometric measurements of

the clearance of exogenously applied serotonin in the rat dentate gyrus. J Neurosci Methods 78:139–150
26. Gobbi M, Frittoli E, Mennini T et al (1992) Releasing activities of d-fenfluramine and fluoxetine on rat hippocampal synaptosomes preloaded with [3H]serotonin. Naunyn Schmiedebergs Arch Pharmacol 345:1–6
27. Gobbi M, Funicello M, Gerstbrein K et al (2008) N,N-Dimethyl-thioamphetamine and methyl-thioamphetamine, two non-neurotoxic substrates of 5-HT transporters, have scant in vitro efficacy for the induction of transporter-mediated 5-HT release and currents. J Neurochem 105:1770–1780
28. Gobbi M, Mennini T, Garattini S (1997) Mechanism of neurotransmitter release induced by amphetamine derivatives: pharmacological and toxicological aspects. Curr Top Pharmacol 3:217–227
29. Rothman RB, Baumann MH (2002) Serotonin releasing agents. Neurochemical, therapeutic and adverse effects. Pharmacol Biochem Behav 71:825–836
30. Rothman RB, Baumann MH (2003) Monoamine transporters and psychostimulant drugs. Eur J Pharmacol 479:23–40
31. Whittaker VP, Michaelson IA, Kirkland RJ (1964) The separation of synaptic vesicles from nerve-ending particles ('synaptosomes'). Biochem J 90:293–303
32. Scholze P, Norregaard L, Singer E et al (2002) The role of zinc ions in reverse transport mediated by monoamine transporters. J Biol Chem 277:21505–21513
33. Eshleman AJ, Henningsen RA, Neve KA et al (1994) Release of dopamine via the human transporter. Mol Pharmacol 45:312–316
34. Wall SC, Gu H, Rudnick G (1995) Biogenic amine flux mediated by cloned transporters stably expressed in cultured cell lines: amphetamine specificity for inhibition and efflux. Mol Pharmacol 47:544–550
35. Pifl C, Agneter E, Drobny H et al (1999) Amphetamine reverses or blocks the operation of the human noradrenaline transporter depending on its concentration: superfusion studies on transfected cells. Neuropharmacology 38:157–165
36. Pifl C, Drobny H, Reither H et al (1995) Mechanism of the dopamine-releasing actions of amphetamine and cocaine: plasmalemmal dopamine transporter versus vesicular monoamine transporter. Mol Pharmacol 47: 368–373
37. Pifl C, Singer EA (1999) Ion dependence of carrier-mediated release in dopamine or norepinephrine transporter-transfected cells questions the hypothesis of facilitated exchange diffusion. Mol Pharmacol 56:1047–1054
38. Seidel S, Singer E, Just H et al (2005) Amphetamines take two to tango: an oligomer-based counter-transport model of neurotransmitter transport explores the amphetamine action. Mol Pharmacol 67:140–151
39. Fog JU, Khoshbouei H, Holy M et al (2006) Calmodulin kinase ii interacts with the dopamine transporter C terminus to regulate amphetamine-induced reverse transport. Neuron 51:417–429
40. Steinkellner T, Montgomery TR, Hofmaier T et al (2015) Amphetamine action at the cocaine- and antidepressant-sensitive serotonin transporter is modulated by alphaCaMKII. J Neurosci 35:8258–8271
41. Steinkellner T, Mus L, Eisenrauch B et al (2014) In vivo amphetamine action is contingent on alphaCaMKII. Neuropsychopharmacology 39: 2681–2693
42. Steinkellner T, Yang JW, Montgomery TR et al (2012) Ca(2+)/calmodulin-dependent protein kinase IIalpha (alphaCaMKII) controls the activity of the dopamine transporter: implications for Angelman syndrome. J Biol Chem 287:29627–29635
43. Rickhag M, Owens WA, Winkler M-T et al (2013) Membrane-permeable C-terminal dopamine transporter peptides attenuate amphetamine-evoked dopamine release. J Biol Chem 288:27534–27544
44. Melikian HE, Buckley KM (1999) Membrane trafficking regulates the activity of the human dopamine transporter. J Neurosci 19: 7699–7710
45. Pifl C, Wolf A, Rebernik P et al (2009) Zinc regulates the dopamine transporter in a membrane potential and chloride dependent manner. Neuropharmacology 56:531–540
46. Foster JD, Yang J-W, Moritz AE et al (2012) Dopamine transporter phosphorylation site threonine 53 regulates substrate reuptake and amphetamine-stimulated efflux. J Biol Chem 287:29702–29712
47. Moritz AE, Foster JD, Gorentla BK et al (2013) Phosphorylation of dopamine transporter serine 7 modulates cocaine analog binding. J Biol Chem 288:20–32
48. Buchmayer F, Schicker K, Steinkellner T et al (2013) Amphetamine actions at the serotonin transporter rely on the availability of phosphatidylinositol-4,5-bisphosphate. Proc Natl Acad Sci U S A 110:11642–11647
49. Hamilton PJ, Belovich AN, Khelashvili G et al (2014) PIP2 regulates psychostimulant behaviors through its interaction with a membrane protein. Nat Chem Biol 10:582–589

50. Scholze P, Freissmuth M, Sitte H (2002) Mutations within an intramembrane leucine heptad repeat disrupt oligomer formation of the rat GABA transporter 1. J Biol Chem 277:43682–43690
51. Chiu CS, Jensen K, Sokolova I et al (2002) Number, density, and surface/cytoplasmic distribution of GABA transporters at presynaptic structures of knock-in mice carrying GABA transporter subtype 1-green fluorescent protein fusions. J Neurosci 22:10251–10266
52. Loland CJ, Norregaard L, Litman T et al (2002) Generation of an activating Zn(2+) switch in the dopamine transporter: mutation of an intracellular tyrosine constitutively alters the conformational equilibrium of the transport cycle. Proc Natl Acad Sci U S A 99: 1683–1688
53. Meinild A, Sitte H, Gether U (2004) Zinc potentiates an uncoupled anion conductance associated with the dopamine transporter. J Biol Chem 279:49671–49679
54. Scholze P, Sitte H, Singer E (2001) Substantial loss of substrate by diffusion during uptake in HEK-293 cells expressing neurotransmitter transporters. Neurosci Lett 309:173–176
55. Rosenauer R, Luf A, Holy M et al (2013) A combined approach using transporter-flux assays and mass spectrometry to examine psychostimulant street drugs of unknown content. ACS Chem Neurosci 4:182–190
56. Baumann MH, Partilla JS, Lehner KR et al (2013) Powerful cocaine-like actions of 3,4-methylenedioxypyrovalerone (MDPV), a principal constituent of psychoactive 'bath salts' products. Neuropsychopharmacology 38:552–562
57. Sitte HH, Freissmuth M (2010) The reverse operation of Na(+)/Cl(−)-coupled neurotransmitter transporters—why amphetamines take two to tango. J Neurochem 112:340–355
58. Mollenhauer HH, Morre DJ, Rowe LD (1990) Alteration of intracellular traffic by monensin; mechanism, specificity and relationship to toxicity. Biochim Biophys Acta 1031:225–246
59. Sitte HH, Scholze P, Schloss P et al (2000) Characterization of carrier-mediated efflux in human embryonic kidney 293 cells stably expressing the rat serotonin transporter: a superfusion study. J Neurochem 74: 1317–1324
60. Chattopadhyay A, Rukmini R, Mukherjee S (1996) Photophysics of a neurotransmitter: ionization and spectroscopic properties of serotonin. Biophys J 71:1952–1960
61. Scholze P, Zwach J, Kattinger A et al (2000) Transporter-mediated release: a superfusion study on human embryonic kidney cells stably expressing the human serotonin transporter. J Pharmacol Exp Ther 293:870–878
62. Koepsell H, Lips K, Volk C (2007) Polyspecific organic cation transporters: structure, function, physiological roles, and biopharmaceutical implications. Pharm Res 24:1227–1251
63. Courousse T, Gautron S (2015) Role of organic cation transporters (OCTs) in the brain. Pharmacol Ther 146:94–103
64. Cui M, Aras R, Christian WV et al (2009) The organic cation transporter-3 is a pivotal modulator of neurodegeneration in the nigrostriatal dopaminergic pathway. Proc Natl Acad Sci U S A 106:8043–8048
65. Iversen LL (1997) The uptake of catechol amines at high perfusion concentrations in the rat isolated heart: a novel catechol amine uptake process. 1964. Br J Pharmacol 120: 267–282, discussion 264–266
66. Vialou V, Balasse L, Callebert J et al (2008) Altered aminergic neurotransmission in the brain of organic cation transporter 3-deficient mice. J Neurochem 106:1471–1482
67. Kristufek D, Rudorfer W, Pifl C et al (2002) Organic cation transporter mRNA and function in the rat superior cervical ganglion. J Physiol 543:117–134
68. Singer EA (1988) Transmitter release from brain slices elicited by single pulses: a powerful method to study presynaptic mechanisms. Trends Pharmacol Sci 9:274–276

Chapter 3

Interrogating the Activity of Ligands at Monoamine Transporters in Rat Brain Synaptosomes

John S. Partilla, Michael H. Baumann, Ann M. Decker, Bruce E. Blough, and Richard B. Rothman

Abstract

The plasma membrane transporters for dopamine (DAT), norepinephrine (NET), and serotonin (SERT) are the main sites of action for therapeutic and abused stimulant drugs. As a means to identify novel medications for stimulant addiction and other psychiatric disorders, we developed in vitro assays in rat brain tissue that can be used to determine structure–activity relationships for test compounds at these monoamine transporters. Uptake inhibition assays measure the ability of drugs to block the transporter-mediated uptake of [^{3}H]neurotransmitters into synaptosomes, whereas release assays measure the ability of drugs to serve as transporter substrates that evoke efflux (i.e., release) of [^{3}H]neurotransmitters from synaptosomes by reverse transport. These assays can be used to rapidly determine the potency of test compounds at DAT, NET, and SERT under similar conditions, establishing the selectivity of drugs across all three transporters. The combined results from uptake and release assays can discriminate whether a compound is a transporter inhibitor or substrate (i.e., releaser). Our assay procedures have been used to characterize the molecular mechanism of action for older amphetamine-type medications and newer transporter ligands with therapeutic potential. The data from these assays can also predict the addictive and neurotoxic properties of abused stimulants. Information provided by these assays continues to provide insight into monoamine transporter structure and function.

Key words Transporter, Synaptosomes, Amphetamine, Stimulants, Neurotransmitter, Uptake, Release

1 Introduction

The plasma membrane neurotransmitter/sodium transporters for dopamine (DAT), norepinephrine (NET), and serotonin (SERT) are the principal sites of action for medications used to treat a range of psychiatric diseases such as depression, anxiety, and attention-deficit hyperactivity disorder [1, 2]. In addition, DAT and NET are implicated in the mechanism of action of addictive stimulants like cocaine and amphetamine [3, 4]. Under normal circumstances, these monoamine transporters serve to translocate previously released neurotransmitter molecules from the extracellular medium

Heinz Bönisch and Harald H. Sitte (eds.), *Neurotransmitter Transporters: Investigative Methods*, Neuromethods, vol. 118,
DOI 10.1007/978-1-4939-3765-3_3,

back into the neuronal cytoplasm, a process known as transporter-mediated "reuptake". Drugs that interact with transporters can be divided into two classes based on their precise molecular mechanism: (1) *inhibitors*—which bind to the neurotransmitter binding site on the extracellular face of the transporter, thereby blocking the reuptake of neurotransmitters from the extracellular medium, and (2) *substrates*—which bind to the transporter and are subsequently translocated through the transporter channel into the neuronal cytoplasm, thereby triggering the efflux of intracellular neurotransmitter molecules (i.e., transporter-mediated release) [5]. Drugs that act as transporter substrates are often called "releasers" because they induce non-exocytotic transporter-mediated neurotransmitter release from neurons.

As a means to identify and characterize new medications for treating stimulant addiction and other psychiatric disorders, we sought to establish in vitro functional assays that could assess the structure–activity relationships for a large library of phenethylamine analogs at DAT, NET, and SERT. We reasoned that these assays should employ a simple and reproducible brain tissue preparation to facilitate high-throughput screening of compounds in a biologically relevant system. Additionally, we designed the assays to allow rapid assessment of potency and efficacy of drugs at all three transporters under similar conditions, and to discriminate whether test compounds act as transporter inhibitors or substrates. It is important to distinguish between transporter inhibitors and substrates because substrate drugs display a number of unique properties: they are translocated into cells along with sodium ions, they induce inward depolarizing sodium currents, and they trigger non-exocytotic release of neurotransmitters by reversing the normal direction of transporter flux (i.e., reverse transport) [6]. Finally, because substrate-type drugs are transported into the neuronal cytoplasm, they can produce intracellular deficits in monoamine neurons such as inhibition of neurotransmitter synthesis leading to long-term neurotransmitter depletions [7, 8].

The purpose of this chapter is to describe straightforward reproducible assays for measuring uptake inhibition and substrate release at DAT, NET, and SERT. All six of the assays use a crude synaptosomal preparation obtained from rat brain and share a common buffer system with only minor differences between assays. Synaptosomes are largely composed of sealed vesicle-filled nerve endings with their plasma membrane leaflets oriented in a manner akin to neurons in vivo [9, 10]. In contrast to assay systems which involve non-neuronal cells transfected with transporter proteins, synaptosomes possess all of the cellular machinery necessary for neurotransmitter synthesis, release, metabolism and reuptake. Our assays are based on the ability to detect the influence of test drugs on transporter-mediated movement of radiolabeled neurotransmitters (i.e., [^{3}H]neurotransmitters) across the synaptosomal plasma membranes.

[^{3}H]-Dopamine ([^{3}H]DA), [^{3}H]-norepinephrine ([^{3}H]NE), and [^{3}H]-serotonin ([^{3}H]5HT) are used as the radiolabeled ligands in the uptake inhibition assays for DAT, NET, and SERT, respectively. [^{3}H]-Methyl-4-phenylpyridinium ([^{3}H]MPP$^+$) is used as the radiolabeled ligand in the release assays for DAT and NET, whereas [^{3}H]5HT is used in the release assay for SERT (*see* **Note 1**). Over the years, these assays have been used to identify transporter drugs of interest that are subsequently tested using more sophisticated or labor-intensive in vivo techniques such as microdialysis, cardiovascular telemetry, intracranial self-stimulation and drug self-administration. Importantly, the pharmacological effects of drugs determined in vivo have been consistent with the drug-transporter relationships described by the uptake and release assays.

2 Materials and Preparation

2.1 Materials

Unless otherwise indicated, all reagents were purchased from Sigma-Aldrich (St. Louis, MO, USA). [^{3}H]DA, [^{3}H]NE, and [^{3}H]5HT were purchased from PerkinElmer Life Sciences (Boston, MA, USA), and [^{3}H]MPP$^+$ was purchased from American Radiolabeled Chemicals (St. Louis, MO, USA). Test compounds were provided by the NIDA IRP Pharmacy (Baltimore, MD, USA), the NIDA Drug Supply Program (Rockville, MD, USA), and Research Triangle Institute (Research Triangle Park, NC, USA).

2.2 Buffers and Assay Selectivity

All uptake inhibition and substrate release assays employ the same two buffers: 0.32 M sucrose that is used for the tissue preparation and comprises 10% of the final reaction volume, and Krebs-phosphate buffer (KPB) that comprises the remaining 90% of the reaction volume. The KPB consists of the following at pH 7.4: 126 mM NaCl, 2.4 mM KCl, 0.5 mM KH_2PO_4, 1.1 mM $CaCl_2$, 0.83 mM $MgCl_2$, 0.5 mM Na_2SO_4, 11.1 mM glucose, 13.7 mM Na_2HPO_4, 1 mg/ml ascorbic acid, and 50 μM pargyline. For substrate release assays, 1 μM reserpine is added to the sucrose and KPB solutions in order to block vesicular uptake of substrates [3]. All buffers are made fresh each day. Assay selectivity is established for a single transporter of interest by adding unlabeled (i.e., nonradioactive) selective transporter blockers to the sucrose solution and KPB to prevent the interaction of radiolabeled ligands and test compounds with competing transporters (see Table 1 and **Note 2**).

2.3 Tissue Preparation

The synaptosomal preparation used in our assays is a slight modification of a procedure originally described over 50 years ago [9]. Male Sprague-Dawley rats are rendered unconscious by CO_2 narcosis, decapitated, and brains are immediately removed. Cerebella are discarded, and paired caudate-putamen are dissected and placed in 10 ml ice cold 0.32 M sucrose (up to three pairs per 10 ml) to be used for DAT studies; the remaining brain tissue is placed in 10 ml ice cold 0.32 M sucrose (one brain per 10 ml) for use in NET and SERT studies. The tissue in sucrose is transferred to a hand-held Teflon-on-glass Potter-Elvehjem tissue grinder and gently homogenized with 12 strokes. The homogenate is centrifuged at 1000 × *g* at 4 °C, and the resulting supernatant is diluted with ice cold 0.32 M sucrose to 17.5 ml for DAT uptake, DAT release, and SERT uptake studies, or 10 ml for NET uptake, NET release, and SERT release studies; the tissue preparation is used immediately (*see* **Note 3**). Electron microscopy reveals that this crude synaptosomal tissue preparation contains intact synaptosomes, as well as myelin fragments, mitochondria, ribosomes, and vesicles [9, 10].

3 Assay Procedures

3.1 Uptake Inhibition

Uptake inhibition assays are based on the exposure of freshly prepared synaptosomes to the appropriate [^{3}H]neurotransmitter of interest (i.e., [^{3}H]DA, [^{3}H]NE, or [^{3}H]5HT). Since synaptosomal plasma membranes are sealed and intact, [^{3}H]neurotransmitter molecules added to the extrasynaptosomal medium are translocated into the intrasynaptosomal compartment by the action of membrane-bound transporter proteins. Under these conditions, transporter-mediated accumulation of [^{3}H]neurotransmitters inside the synaptosomes is linearly proportional to time (data not shown). During a typical uptake inhibition experiment, a solution containing [^{3}H]neurotransmitter and test drug is mixed with synaptosomes; this mixture is allowed to incubate for a specific period of time, within the linear portion of the [^{3}H]neurotransmitter accumulation time course. To stop the incubation, the mixture is rapidly filtered, thereby trapping accumulated intrasynaptosomal [^{3}H]neurotransmitter molecules on the filter (Whatman GF/B fiberglass filters, Brandel, Gaithersburg, MD, USA) while remaining "free" extrasynaptosomal [^{3}H]neurotransmitters are removed to waste. In this manner, the filtration process captures the [^{3}H] neurotransmitters taken up by synaptosomes during the incubation. Retained radioactivity is then quantified by liquid scintillation counting and is proportional to the quantity of synaptosomes used in the assay. If test drugs interact with monoamine transporters, the accumulation of [^{3}H]neurotransmitter inside of the synaptosomes (i.e., uptake) is reduced because the test drug and [^{3}H]neurotransmitter molecules compete for the same binding site on the transporter protein. This degree of "uptake inhibition" is proportional to the concentration of test drug; the concentration of test drug required to inhibit uptake by 50% (IC_{50}) defines the potency of the uptake inhibitor for the transporter.

We first used this technique to examine the transporter selectivity of a homolog of the DAT blocker GBR12935, LR1111 [11], and continue to use the method with only minor modifications [12]. Uptake inhibition assays are conducted in triplicate at 25 °C (DAT and SERT) or 37 °C (NET) in polystyrene test tubes (12×75 mm). Assays are initiated by the addition of 100 μl of freshly prepared synaptosomes to 900 μl KPB that contains the appropriate [^{3}H]neurotransmitter, selective blockers (see Table 1), and test drug. With regard to radiolabeled transmitter concentrations, we use final concentrations of 5 nM [^{3}H]DA, 10 nM [^{3}H]NE and 5 nM [^{3}H]5HT for the DAT, NET, and SERT uptake inhibition assays, respectively. Test drug solutions are prepared in KPB containing 1 mg/ml bovine serum albumin over a range of eight different doses in order to construct an inhibition curve that establishes an IC_{50} value for the test drug. Nonspecific uptake is measured by incubating synaptosomes in the presence of a large excess (1 μM) of the nonselective uptake inhibitor indatraline. The reactions are stopped after 15 min (DAT), 10 min (NET), or 30 min (SERT) by rapid vacuum filtration with a cell harvester (Brandel, Gaithersburg, MD, USA) over GF/B filter paper presoaked in wash buffer maintained at 25 °C (10 mM Tris-HCl, pH 7.4/150 mM NaCl). Filters are rinsed with 6 ml wash buffer, and retained tritium is quantified by a MicroBeta 2 liquid scintillation counter (PerkinElmer, Boston, MA, USA) after overnight extraction in 0.6 mL of liquid scintillation cocktail (Cytoscint, MP Biomedicals, Santa Ana, CA, USA).

In order to describe the method for calculating the uptake IC_{50}, the following definitions are necessary:

Total Uptake (TU) = cpm accumulated in the absence of any drug.

Nonspecific Uptake (NS) = cpm accumulated in the presence of total transporter blockade using1 μM indatraline.

Table 1
Uptake inhibitors used to establish assay selectivity

Assay	Desipramine (blocks NET)	Citalopram (blocks SERT)	GBR 12935 (blocks DAT)	Nomifensine (blocks NET)
DAT uptake[a]				
SERT uptake			50 nM	100 nM
NET uptake[b]			50 nM	
DAT release	100 nM	100 nM		
SERT release			50 nM	100 nM
NET release		100 nM	50 nM	

Selective uptake inhibitors are added in the indicated concentrations to both the sucrose solution and KPB
[a]DAT uptake selectivity is established by surgical isolation of the caudate-putamen, a region so enriched in DAT that measurable uptake of [^{3}H]DA by NET or SERT does not occur; selective blockade of NET or SERT is not required
[b]SERT affinity for [^{3}H]NE is so weak that SERT blockade is not required

Maximal uptake inhibition (MU) = TU − NS.

Specific uptake inhibition (SU) = (cpm in the presence of test drug) − NS.

% uptake inhibition = 100 − SU/MU × 100.

The data from three experiments, expressed as % uptake inhibition, are then fit to a dose-response curve equation: $Y = I_{max} \times ([D]/([D] + IC_{50}))$ for the best fit estimates of the I_{max} and IC_{50}, where [D] is the concentration of test drug. Over the years, we have used a number of different commercially available software programs for data analysis and curve fitting, including MLAB-PC (Civilized Software, Silver Spring, MD, USA), KaleidaGraph (Synergy Software, Reading, PA, USA), and Prism (GraphPad Software, La Jolla, CA, USA).

Similar to uptake inhibitors, transporter substrates will compete with [^{3}H]neurotransmitter molecules to occupy transporter binding sites during uptake inhibition assays. Additionally, substrate drugs will evoke efflux of any accumulated transmitter. The combined effects of substrate drugs will reduce the amount of [^{3}H]neurotransmitter accumulated into synaptosomes in the uptake inhibition assays. Thus, uptake inhibition assays cannot discriminate whether transporter drugs are inhibitors or substrates. Nevertheless, our initial experiments demonstrated that a number of known transporter substrates (e.g., amphetamine-related compounds) were generally three- to tenfold more potent in substrate release assays when compared to their potency in the uptake inhibition assays (see Table 2) [3]. This pattern has persisted with nearly all substrate drugs that we have tested in the uptake inhibition assays [13, 14].

3.2 Substrate Release

Using the uptake inhibition assay conditions described above, synaptosomes will accumulate [^{3}H]neurotransmitter molecules in a biphasic manner, with an initial linear burst followed by a slower phase that does not plateau. The initial phase of [^{3}H]neurotransmitter uptake is mediated by the action of plasma membrane monoamine transporters (i.e., DAT, NET and SERT), whereas the later phase is related to the uptake of neurotransmitters into synaptic vesicles by the action of the vesicular monoamine transporter-2 (VMAT-2). Inclusion of 1 μM reserpine, an irreversible blocker of VMAT-2, in the sucrose and KPB solutions allows for the steady state accumulation of [^{3}H]neurotransmitter molecules into the intrasynaptosomal compartment within 60 min, by preventing the uptake of [^{3}H]neurotransmitter and substrate drugs into synaptic vesicles that are present in our tissue preparation [3]. Thus, the inclusion of reserpine in the sucrose and KPB removes any effects of VMAT-2 from the assay system.

The substrate release assay is based on the efflux of previously accumulated (i.e., preloaded) synaptosomal [^{3}H]neurotransmitter by exposure to test substrates, via a transporter-mediated exchange

Pharmacological profile of selected agents in transporter release and uptake inhibition assays[a]

Drug	DAT release	DAT uptake	NET release	NET uptake	SERT release	SERT uptake
	EC_{50} (nM ± SD)	IC_{50} (nM ± SD)	EC_{50} (nM ± SD)	IC_{50} (nM ± SD)	EC_{50} (nM ± SD)	IC_{50} (nM ± SD)
Aminorex	49.4 ± 7.5	216 ± 7	26.4 ± 2.8	54.5 ± 4.8	193 ± 23	1244 ± 106
Phentermine	262 ± 21	1580 ± 80	39.4 ± 6.6	244 ± 15	3511 ± 253	13,900 ± 510
(+)-Amphetamine	24.8 ± 3.5	34.0 ± 6.0	7.07 ± 0.95	38.9 ± 1.8	1765 ± 94	3830 ± 170
(−)-Methamph	416 ± 20	4840 ± 178	28.5 ± 2.5	234 ± 14	4640 ± 243	14,000 ± 644
(+)-Methamph	24.5 ± 2.1	114 ± 11	12.3 ± 0.7	48.0 ± 5.1	736 ± 45	2137 ± 98
(+)-Fenfluramine	>10,000	22,000 ± 1100	302 ± 20	1290 ± 152	51.7 ± 6.1	150 ± 5
(±)-Fenfluramine	>10,000	23,700 ± 1300	739 ± 57	1987 ± 205	79.3 ± 11.5	269 ± 7
(−)-Ephedrine	1350 ± 12	4398 ± 213	72.4 ± 10.2	225 ± 36	>10,000	>50,000
Tyramine	119 ± 11	106 ± 6	40.6 ± 3.5	72.5 ± 5.0	2775 ± 234	1556 ± 95
(±)-MDMA	376 ± 16	1572 ± 59	77.4 ± 3.4	462 ± 18	56.6 ± 2.1	238 ± 13
Norepinephrine	869 ± 51	357 ± 27	164 ± 13	63.9 ± 1.6	>10,000	>50,000
Dopamine	86.9 ± 9.7	38.3 ± 1.6	66.2 ± 5.4	40.3 ± 4.4	>10,000	6489 ± 200
5-HT	>10,000	2703 ± 79	>10,000	3013 ± 266	44.4 ± 5.3	16.7 ± 0.9
GBR12935	>10,000	3.70 ± 0.40	>10,000	277 ± 23	>10,000	289 ± 29
Mazindol	>10,000	25.9 ± 0.6	>10,000	2.88 ± 0.17	>10,000	272 ± 11
Desipramine	>10,000	5946 ± 193	>10,000	8.32 ± 1.19	>10,000	350 ± 13
Fluoxetine	>10,000	>5000	>10,000	688 ± 39	>10,000	9.58 ± 0.88
Citalopram	>10,000	20,485 ± 923	>10,000	4332 ± 295	>10,000	2.40 ± 0.09
RTI-55	>10,000	0.83 ± 0.09	>10,000	5.89 ± 0.53	>10,000	1.00 ± 0.03
Indatraline	2810 ± 777	1.90 ± 0.05	>10,000	12.6 ± 0.5	>10,000	3.10 ± 0.09

Each value is the mean ± SD of three experiments

[a]The data in this table were reported in ref. [3]

process thought to involve the reversal of normal transporter flux (i.e., "reverse" transport) [15, 16]. Substrate-type drugs will deplete [^{3}H]neurotransmitter from synaptosomes via this reverse transport mechanism in a concentration-dependent manner; the amount of test drug required to release 50% of the preloaded [^{3}H] neurotransmitter (EC_{50}) defines the potency of that substrate for the transporter. Similar to the uptake inhibition assay, transporter selectivity in the substrate release assay is achieved by inclusion of unlabeled selective uptake inhibitors in the sucrose solution and KPB to prevent the interaction of [^{3}H]neurotransmitters and test compounds with competing transporters (see Table 1). For the SERT release assay, [^{3}H]5HT is used as the radiolabeled neurotransmitter. [^{3}H]MPP^+ is used for the DAT and NET release assays because it is less diffusible across cell membranes and produces a better signal-to-noise ratio than [^{3}H]DA and [^{3}H]NE.

Prior to the release assay, synaptosomes must be preloaded with [^{3}H]neurotransmitters. To accomplish this step, synaptosomes are incubated with the appropriate [^{3}H]neurotransmitter in the presence of selective uptake inhibitors (see Table 1) in KPB containing 1 μM reserpine. The incubation is allowed to reach equilibrium (1 h at 25 °C). To initiate the release reaction, 850 μl of synaptosomes preloaded with [^{3}H]neurotransmitter are added to 12×75 mm polystyrene test tubes that contain 150 μl of test drug that has been diluted in KPB containing 1 mg/ml bovine serum albumin. Each test condition in the release assay is run in triplicate. After 30 min (DAT, NET) or 5 min (SERT), the release reactions are stopped using a cell harvester (Brandel), by rapid vacuum filtration over GF/B filter paper presoaked in wash buffer maintained at 25 °C (10 mM Tris-HCl, pH 7.4, 150 mM NaCl). For the NET release assay, the wash buffer contains 2% polyethylenimine to minimize nonspecific adsorption of [^{3}H]MPP^+ to the filter paper. Filters are rinsed with 6 ml wash buffer, and retained tritium is quantified by a MicroBeta 2 liquid scintillation counter after overnight extraction in 0.6 mL of Cytoscint. For the release assay, it is important to note that the amount of tritium retained is inversely proportional to the extent of release from synaptosomes; that is, a lower amount of retained tritium reflects a higher degree of transporter-mediated release.

In order to describe the method for calculating the release EC_{50}, the following definitions are necessary:

Basal "Leak" Release (BLR) = cpm in the absence of any drug.

Total Evoked Release (TER) = cpm in the presence of saturating concentrations of the nonspecific releaser, tyramine—10 μM tyramine for DAT and NET assays, or 100 μM tyramine for SERT assays.

Maximal release (MR) = BLR − TER.

Specific release (SR) = (cpm in the presence of test drug) − TER.

% maximal release = 100 − SR/MR × 100.

The data from three experiments, expressed as % maximal release, are then fit to a dose-response curve equation: $Y = E_{max} \times ([D]/([D] + EC_{50}))$ for the best fit estimates of the E_{max} and EC_{50}, where [D] is the concentration of test drug. Data analysis and curve fitting can be accomplished using the same software programs noted above for the uptake inhibition assays.

In the substrate release assays, test drugs are not exposed to synaptosomes until the synaptosomes have been preloaded with [^{3}H]neurotransmitter. Subsequent exposure of preloaded synaptosomes to test drugs that are substrates provokes the depletion (i.e., release) of [^{3}H]neurotransmitter from the synaptosomes, via transporter-mediated exchange, in a concentration-dependent manner [17, 18]. Importantly, uptake inhibitors show no or minimal activity in substrate release assays because they are not transported by DAT, NET, and SERT and cannot evoke release of [^{3}H] neurotransmitter from preloaded synaptosomes (see Table 2). Therefore, unlike the uptake inhibition assays, the substrate release assays are able to discriminate between drugs that are uptake inhibitors and drugs that are substrates at transporters.

4 Summary

The uptake inhibition and substrate release assays provide straightforward, relatively high-throughput methods for assessing the interaction of drugs with the monoamine transporter proteins for dopamine, norepinephrine and serotonin. We use a crude synaptosomal tissue preparation that is simple to generate, yet provides highly reproducible results. These assays allow for the determination of relative potencies of drugs at DAT, NET, and SERT, and are able to distinguish drugs that are transporter uptake inhibitors from those that are transporter substrates. Data from the combined assay results are used to identify drugs of interest that can be further evaluated by in vivo techniques such as microdialysis, cardiovascular telemetry, intracranial self-stimulation and drug self-administration.

We have used the uptake inhibition and substrate release assays to reexamine the mechanism of action for anorectic medications that were marketed in the 1970s, such as fenfluramine and diethylpropion. For fenfluramine, we found that the parent compound and its *N*-dealkylated metabolite, norfenfluramine, display potent substrate activity at NET, as well as their established substrate activity at SERT [19]. The SERT substrate activity of fenfluramine and norfenfluramine has been linked to adverse effects such as primary pulmonary hypertension. In the case of diethylpropion, it was discovered that the parent compound is devoid of transporter activity while the *N*-dealkylated metabolite, *N*-ethylcathinone, is active as a DAT inhibitor and SERT substrate [20]. Thus, diethylpropion is a prodrug which requires hepatic biotransformation to *N*-ethylcathinone to exert its therapeutic actions.

Testing large libraries of compounds in our assays has led to the identification of ligands with novel profiles of action at monoamine transporters. For example, most transporter inhibitors competitively block the uptake of [^{3}H]neurotransmitter by binding to the orthosteric site on the transporter protein, but we have discovered quinazolinamine analogs which noncompetitively inhibit uptake at DAT, NET, and SERT [12]. These allosteric modulators potently inhibit the uptake of [^{3}H]neurotransmitter without substantially affecting transporter ligand binding or amphetamine-induced release. Such results suggest that these compounds interact with monoamine transporters in a manner which selectively interrupts the uptake portion of the transport cycle. In other studies, we have identified unique transporter substrates which evoke only 50–70% of the maximal release response, indicative of "partial" releasing ability [21]. The molecular mechanism responsible for partial release is unknown and under investigation, but these compounds blur the distinction between "pure" uptake inhibitors and "pure" substrate-type releasers. From a drug development perspective, allosteric modulators and partial substrates at monoamine transporters may represent promising new leads for medication discovery.

Recently, we have used the uptake inhibition and substrate release assays to determine the molecular mechanism of action for synthetic cathinone compounds that are the active ingredients in psychoactive "bath salts" products [14, 22]. These products are used as substitutes for more traditional stimulant drugs of abuse such as cocaine and amphetamine. In initial experiments, we demonstrated that 3,4-methylenedioxypyrovalerone (MDPV) is a potent uptake inhibitor at DAT and NET, whereas 4-methylmethcathinone and 3,4-methylenedioxymethcathinone are nonselective transporter substrates [14, 22]. Subsequent structure–activity studies revealed that other cathinone analogs which possess an *N*-containing pyrrolidine ring, similar to MDPV, act as catecholamine-selective inhibitors, probably because these compounds are sterically too large to permeate the transporter channel as substrates [23, 24]. Phenyl ring-substitutions can markedly influence the transporter selectivity for cathinone compounds, such that bulky *para* substituents engender increased potency towards SERT versus DAT [25, 26]. A shift in selectivity towards SERT over DAT reduces the abuse liability of cathinones, similar to the findings with ring-substituted amphetamines [7, 25]. Results from our uptake and release assays have been used to explore the interaction of cathinone compounds with transporters at the molecular level, using models of DAT and SERT based on the reported crystal structures of bacterial LeuT and *Drosophila* DAT proteins [23, 26]. It is hoped that new research findings generated from our in vitro assays will continue to provide insights into the structure and function of monoamine transporter proteins.

Notes

1. [^{3}H]DA, [^{3}H]NE, [^{3}H]5HT, and [^{3}H]MPP$^+$ are stored at −80 °C in their original source vials and are thawed on the day of assay at room temperature. Every effort is made to minimize the amount of time that [^{3}H]neurotransmitters in their source vials are in the thawed liquid state, in order to minimize chemical degradation. Aliquoting to eliminate degradation related to repeated freeze–thaw cycles does not improve the stability of the radioligands, as measured by the observed signal-to-noise ratio of each assay.
2. All buffers are freshly made each day. To simplify the procedure, the following KPB components are combined in a partial 10× stock solution (10× KPB) that is aliquoted, stored frozen, and used as needed: NaCl, KCl, KH_2PO_4, $CaCl_2$, $MgCl_2$, and Na_2SO_4. To prepare KPB, a 30 ml aliquot of 10× KPB is thawed and diluted to 300 ml with water. Ascorbic acid (300 mg), 0.585 g Na_2HPO_4, 0.6 g dextrose, and 3 mg pargyline are added, and the pH is adjusted to 7.4 with drop wise 2 N NaOH. Appropriate selective blockers (and reserpine for release assays) are then added (see Table 1). Seventy-five milliliters of the KPB is used for drug dilutions, and bovine serum albumin is added to this solution to yield a final concentration of 1 mg/ml.
3. Synaptosomes are prepared each day from freshly excised rat brain and are used immediately after the centrifugation step to initiate the uptake inhibition and substrate release assays. Under these conditions the synaptosomes are able to maintain [^{3}H] neurotransmitter equilibrium for up to 4 h.

References

1. Gorman JM, Kent JM (1999) SSRIs and SMRIs: broad spectrum of efficacy beyond major depression. J Clin Psychiatry 60(Suppl 4): 33–38
2. Iversen L (2006) Neurotransmitter transporters and their impact on the development of psychopharmacology. Br J Pharmacol 147(Suppl 1):S82–S88
3. Rothman RB et al (2001) Amphetamine-type central nervous system stimulants release norepinephrine more potently than they release dopamine and serotonin. Synapse 39:32–41
4. Howell LL, Kimmel HL (2008) Monoamine transporters and psychostimulant addiction. Biochem Pharmacol 75:196–217
5. Rothman RB, Baumann MH (2003) Monoamine transporters and psychostimulant drugs. Eur J Pharmacol 479:23–40
6. Sitte HH, Freissmuth M (2015) Amphetamines, new psychoactive drugs and the monoamine transporter cycle. Trends Pharmacol Sci 36:41–50
7. Baumann MH, Wang X, Rothman RB (2007) 3,4-Methylenedioxymethamphetamine (MDMA) neurotoxicity in rats: a reappraisal of past and present findings. Psychopharmacology (Berl) 189:407–424
8. Fleckenstein AE et al (2007) New insights into the mechanism of action of amphetamines. Annu Rev Pharmacol Toxicol 47:681–698

9. Gray EG, Whittaker VP (1962) The isolation of nerve endings from brain: an electron-microscopic study of cell fragments derived by homogenization and centrifugation. J Anat 96:79–88
10. Wilhelm BG et al (2014) Composition of isolated synaptic boutons reveals the amounts of vesicle trafficking proteins. Science 344:1023–1027
11. Rothman RB et al (1993) Identification of a GBR12935 homolog, LR1111, which is over 4,000-fold selective for the dopamine transporter, relative to serotonin and norepinephrine transporters. Synapse 14:34–39
12. Rothman RB et al (2015) Studies of the biogenic amine transporters 15. Identification of novel allosteric dopamine transporter ligands with nanomolar potency. J Pharmacol Exp Ther 353(3):529–538
13. Rothman RB et al (2000) Methamphetamine dependence: medication development efforts based on the dual deficit model of stimulant addiction. Ann N Y Acad Sci 914:71–81
14. Baumann MH et al (2013) Powerful cocaine-like actions of 3,4-methylenedioxypyrovalerone (MDPV), a principal constituent of psychoactive 'bath salts' products. Neuropsychopharmacology 38(4):552–562
15. Rudnick G, Clark J (1993) From synapse to vesicle: the reuptake and storage of biogenic amine neurotransmitters. Biochim Biophys Acta 1144:249–263
16. Rothman RB, Baumann MH (2006) Therapeutic potential of monoamine transporter substrates. Curr Top Med Chem 6: 1845–1859
17. Rothman RB et al (2003) In vitro characterization of ephedrine-related stereoisomers at biogenic amine transporters and the receptorome reveals selective actions as norepinephrine transporter substrates. J Pharmacol Exp Ther 307:138–145
18. Scholze P et al (2000) Transporter-mediated release: a superfusion study on human embryonic kidney cells stably expressing the human serotonin transporter. J Pharmacol Exp Ther 293:870–878
19. Rothman RB et al (2003) (+)-Fenfluramine and its major metabolite, (+)-norfenfluramine, are potent substrates for norepinephrine transporters. J Pharmacol Exp Ther 305:1191–1199
20. Yu H et al (2000) Uptake and release effects of diethylpropion and its metabolites with biogenic amine transporters. Bioorg Med Chem 8:2689–2692
21. Rothman RB et al (2012) Studies of the biogenic amine transporters. 14. Identification of low-efficacy "partial" substrates for the biogenic amine transporters. J Pharmacol Exp Ther 341(1):251–262
22. Baumann MH et al (2012) The designer methcathinone analogs, mephedrone and methylone, are substrates for monoamine transporters in the brain. Neuropsychopharmacology 37(5):1192–1203
23. Saha K et al (2015) 'Second-generation' mephedrone analogs, 4-MEC and 4-MePPP, differentially affect monoamine transporter function. Neuropsychopharmacology 40(6):1321–1331
24. Marusich JA et al (2014) Pharmacology of novel synthetic stimulants structurally related to the "bath salts" constituent 3,4-methylenedioxypyrovalerone (MDPV). Neuropharmacology 87:206–213
25. Bonano JS et al (2015) Quantitative structure-activity relationship analysis of the pharmacology of para-substituted methcathinone analogues. Br J Pharmacol 172(10):2433–2444
26. Sakloth F et al (2015) Steric parameters, molecular modeling and hydropathic interaction analysis of the pharmacology of para-substituted methcathinone analogues. Br J Pharmacol 172(9):2210–2218

Chapter 4

Using High-Speed Chronoamperometry to Measure Biogenic Amine Release and Uptake In Vivo

Lynette C. Daws, W. Anthony Owens, and Glenn M. Toney

Abstract

Here, we describe the method of high-speed chronoamperometry and its application for measuring release and clearance of biogenic amine neurotransmitters (serotonin, dopamine, and norepinephrine) in the intact and living mammalian brain. Chronoamperometry belongs to a family of electrochemical techniques collectively known as voltammetry, the only techniques currently available for "real-time" measurement of neurotransmitter transporter activity in vivo. Because of the small size of recording electrodes (<30 μm) and the relatively rapid sampling rate (sub-second), these techniques can be used to quantify release and clearance kinetics for biogenic amines in discrete brain regions. Chronoamperometry has been effectively used to study the impact of drugs, various environmental influences (e.g. stress), the estrous cycle, and age, among other stimuli, on the function of biogenic amine transporters in vivo.

A major part of performing high-speed chronoamperometry is the preparatory work, including fabricating and calibrating carbon fiber electrodes, creating electrode-micropipette assemblies and stereotaxically implanting them in brain. Details for all steps are provided here, including how to histologically verify electrode placement at the conclusion of recordings. Chronoamperometry provides a unique window to "view" biogenic amine transporter function in the living animal.

Key words Chronoamperometry, In vivo electrochemistry, Carbon fiber electrode, Serotonin, Dopamine, Norepinephrine, Biogenic amine transporters

1 Introduction

Biogenic amine neurotransmitters, including dopamine (DA), norepinephrine (NE), and serotonin (5-HT), play pivotal roles in many complex behaviors and physiological functions. These include regulation of mood, reward, arousal, sleep, thermoregulation, appetitive behavior, immune response, cardiovascular, and gut function, to name a few. Not surprisingly, dysregulation of biogenic amine neurotransmission can have severe health consequences. Of fundamental importance for advanced studies of biogenic aminergic neurotransmission is developing tools to accurately quantify changes in neurotransmitter concentration in living animals in a time frame that captures the dynamics of

Heinz Bönisch and Harald H. Sitte (eds.), *Neurotransmitter Transporters: Investigative Methods*, Neuromethods, vol. 118, DOI 10.1007/978-1-4939-3765-3_4,

neurotransmitter release and uptake. Traditionally, measures of biogenic amine release and uptake have been made using in vitro approaches, which measure efflux and uptake of radiolabeled transmitter or other cations such as 1-methyl-4-phenylpyridinium (MPP^+) and 4-(4-(dimethylamino)styryl)-*N*-methylpyridinium iodide (ASP^+). These in vitro approaches and their increasing level of sophistication have yielded tremendous insight into the regulation of biogenic amine neurotransmission and in particular, transporter function. However, by their very nature, ex vivo tissue preparations, or cultured cells expressing transmitters or transporters of interest do not provide insight into the kinetics of biogenic amine neurotransmission in the living, intact animal. Such insight is essential if we are to fully translate the relevance of in vitro observations to systems level function.

To this end, the age of in vivo electrochemistry, pioneered by Ralph Adams in the 1970s [1], has allowed for rapid measurement of biogenic amine neurotransmission in vivo. Electrochemical techniques include various forms of cyclic voltammetry and high-speed chronoamperometry. Here, we describe the method of high-speed chronoamperometry to measure biogenic amine release and uptake in vivo. The principle of high-speed chronoamperometric recording is based on methods originally described by Cottrell [2]. We have adopted and modified methods developed by the Gerhardt group, who extended work begun in Ralph Adams' laboratory to develop electronics and digital data processing to allow for more rapid measurements (millisecond to second time scale) of dopamine, glutamate, and other neurotransmitters. This temporal resolution, combined with the spatial resolution afforded by small-diameter carbon-fiber recording electrodes (10–30 μm), has allowed us to measure the rapid kinetics of neurotransmitter release and uptake in discrete regions of brain in the living animal. In the following sections, we describe procedures that we have adopted and refined to measure release and uptake of biogenic amines in vivo.

2 Materials and Methods

2.1 Major Pieces of Specialized Equipment and Materials Required for High-Speed Chronoamperometry and Suggested Vendors

- Either a FAST-12 or FAST-16 system from Quanteon (Nicholasville, KY, USA; http://www.quanteon.cc/home), or a self-assembled system by purchasing component parts (Axon Instruments: http://www.moleculardevices.com/systems/axon-conventional-patch-clamp; Heka Elektronik: http://www.heka.com/products/products_main.html#physiol_ampero) that include a relatively high-impedance headstage coupled to a stable potentiostat (current-to-voltage converter) along with data acquisition and analysis software (*see* **Note 1**).
- Computer (recommended minimum specifications: 2.5 GHz processor; 2 GB RAM; 100 GB Hard Drive; 1280 × 1024

flat-panel 17″ monitor. This is for current FAST 16 systems in 2014).

- Dual/Multi-channel Picospritzer III (Parker Corp., Cleveland, OH, USA).
- Nitrogen gas tank (60″ compressed nitrogen).
- Dissecting microscope (EMZ Series stereoscope from Meiji (Itasca, IL, USA); 0.7–4.5× zoom with eye-piece reticule (10× magnification)) fitted to a boom stand (Diagnostic Instruments SMS25, Sterling Heights, MI, USA).
- Fiberoptic lamp (Kent Scientific Corp., Torrington, CT, 06790, USA).
- Micromanipulator (Stoelting, MM33 tilt-base, roller bearing micromanipulator, Wood Dale, IL, USA).
- Microdrive (Narishige MO-10 hydraulic microdrive, East Meadow, NY, USA).
- Nafion-coated, carbon-fiber recording electrodes (see Sect. 2.2).
- Multi-barreled micropipettes (4-Barrel (14-15-L); 7-Barrel (17-15-L): [1.5 mm OD] with tip modification (9–12 μm) by FHC Inc., Bowdoin, ME, USA).
- Sterotaxic frame (for mouse or rat; David Kopf or Stoelting).
- Circulating water heater pad (Gaymar, TP500 water-heated hypothermia unit, Orchard Park, NY, USA).
- Rectal thermometer (Model 4600 Precision thermometer with rectal probe for rodents by YSI, Dayton, OH, USA).
- Blood oximeter (MouseOx small animal oximeter from Starr Life Sciences, Allison Park, PA, USA).

2.2 Materials for General Operation

- Kerr™ dental sticky wax (Sybron Dental Specialties, Orange, CA, USA).
- Polyethylene tubing (Intramedic, non-radiopaque polyethylene tubing (PE10; PE50), Becton Dickinson, Parsipanny, NY, USA).
- Access to fume-hood (for preparing anesthetic).
- Stainless steel trachea tubes. For rats: OD = 0.083″; ID = 0.063″; cut to ~2 in. in length. Mice: OD = 0.05″; ID = 0.033″; cut to ~1/2 in. length (stainless steel, hypodermic tubing Small Parts Inc., Miami Lakes, FL, USA).
- Chemicals for standard buffers (e.g. phosphate buffered saline [PBS], artificial cerebral spinal fluid [aCSF]). For 1 l 0.1 M PBS, $Na_2HPO_4{\cdot}7H_2O$ (20.98 g), $Na_2HPO_4{\cdot}H_2O$ (2.48 g), NaCl (9 g); start with 800–900 ml H_2O. Stirring, add the compounds one at a time and wait for each to completely dissolve. Then add H_2O to achieve a final volume of 1 l. Final pH should be 7.2–7.4.

For 2 . L aCSF, 126 mM (NaCl); 2.9 mM (KCl); 1.5 mM (MgCl); 2.5 mM (CaCl); 1.4 mM (NaH_2PO_4); 1.4 mM ($NaHCO_3$); 6.846 g sucrose; q.s. to 2 L with ddH_2O. Must be oxygenated daily (95 % O_2/5 % CO_2) for ~10 min. pH adjusted to 7.25–7.40 (0.1 M HCl; 0.1 M NaOH). aCSF stock is reusable for 1–2 months, provided no bacterial growth or contamination.

- Neurotransmitters and drugs of interest.

  The use of these is described in the following sections.

2.3 Nafion-Coated Carbon Fiber Electrodes (CFE)

Nafion is a perfluronated ion exchange resin that excludes the passage of anions to the CFE [3, 4]. It has been used with great success to increase the selectivity of CFEs for cations such as 5-HT, DA, and NE, and prevents oxidation of anionic "contaminants" such as the metabolites of these neurotransmitters as well as uric acid and ascorbic acid, which are present in the brain in much higher concentrations than the biogenic amines and have oxidation potentials that overlap with those of the biogenic amines [5–8]. If oxidation of anionic species is not prevented by Nafion, then interpreting which species is being oxidized at the implanted CFE becomes extremely difficult, if not impossible.

Nafion-coated CFEs can be purchased pre-fabricated from Quanteon (http://www.ukycenmet.com/carbon-fiber-for-catecholamine/) (*see* **Note 2**), or made in-house.

2.3.1 If Making In-House, You Will Need the Following Materials

- Fused silica glass tubing—Sodalime (4 mm O.D.; 0.7 mm wall thickness) for working electrodes (Schott Rohrglas #26005, Glass Warehouse, Millville, NJ, USA).
- Carbon fibers (30 μm carbon monofilament, AVCO Specialty Materials, Subsidiary of Textron Inc., Lowell, MA, USA; http://www.specmaterials.com/index.htm).
- Graphite-epoxy paste (carbon epoxy; Item #7335, World Precision Instruments (WPI), Sarasota, FL, USA; www.wpi-inc.com).
- Epoxy (Epoxylite #6001-M, The Epoxylite Corporation, Westerville, OH, USA).
- Copper wire (Pare Wire 28 Gauge monofilament, non-tarnish copper, 300 ft, Paramount Wire Co., East Orange, NJ, USA; www.**parawire**.com).
- Gold pin connectors (from any electronics store).
- Nafion-5 % w/v solution (Sigma-Aldrich).
- Silver chloride (AgCl) reference electrode (e.g. Model # RE-5B; MF-2079 p*k* of 3, Bioanalytical Systems, West Lafayette, IN, USA).

Steps to make electrodes in-house follows (see also [9] and references therein).

2.3.2 There Are Four Main Phases in the Fabrication of CFEs

Phase 1: Cutting and pulling fused silica glass tubing.

Phase 2: Inserting carbon fiber, packing with epoxy and attaching copper wire.

Phase 3: Cutting tip of carbon fiber and checking for electrical continuity.

Phase 4: Cleaning and Nafion coating.

Phase 1: Cutting and pulling glass capillary tubing (fused silica tubing, Schott)

- Using a file, score and snap silica tubing into 1½ to 2 in. lengths.
- Using a vertical glass puller (e.g. Narishige PE2 vertical glass puller, East Meadow, NY, USA) pull silica tubing to a taper approximately 3 cm long.

Phase 2: Inserting carbon fiber (30 μm diameter, Specialty Materials)

- Fill a 3 or 5 ml syringe with epoxylite and attach a 25 g needle.
- Cut a 5–6 cm length of carbon fiber and insert into silica tubing.
- It will be necessary to use a Craftsman™ rotary grinding cut-off wheel (or similar) fitted to a Dremel™ tool (or similar tool) to open the tip of the pulled silica tubing in order for the carbon fiber to pass through.
- Insert needle with epoxylite into the non-tapered end of the silica tubing and inject epoxylite. Be certain not to fill all the way to the pulled tip of the glass capillary. If epoxylite gets on to the tip of the electrode, the electrode will not work. Aspirate excess epoxylite into syringe.
- Set in petri dish and bake in oven (e.g. Precision™ compact oven by Thermo-Fisher Scientific, Waltham, MA, USA or similar) overnight at 120 °C.

Packing electrode with carbon epoxy

- Cut copper wire into 11–12 cm long strips.
- Remove insulation from both ends of wire strands using wire strippers.
- Pack carbon epoxy into non-tapered end of silica tubing and compress using a toothpick.
- Break off non-tapered part of silica tubing with small wire cutters leaving only the small tapered end (~3 cm long).

- Insert wire ~0.5 cm into epoxy. Make sure to leave sufficient wire (~10.5 cm) extending from the electrode so that you will be able to connect the electrode to the headstage once the electrode is implanted into brain.
- Set aside to bake overnight at 100 °C.

Phase 3: Cutting CFE and checking for electrical continuity

- Under microscope, cut tip of the carbon fiber with a number 10 stainless steel surgical blade so that it protrudes ~150 μm from glass. The length of carbon fiber protruding can be varied. For example, increasing the length increases the recording surface area thereby increasing the sensitivity of the electrode to detect neurotransmitter, but at the expense of reducing spatial resolution of recording area.
- Solder gold pin ampule to copper wire.
- To check for electrical continuity, fill a beaker with PBS, lower the AgCl reference electrode into the solution. Lower CFE into solution. Apply a potential of +0.55 V to the CFE. If a current flows (~25 nA), your electrode is good. If not, try re-soldering the gold-pin connector to the wire. If still no connectivity, discard the electrode. Inadequate packing of epoxy, or epoxy covering the exposed carbon fiber tip, likely occurred.
- Place working CFEs in plastic storage box, such that CFEs are secure and will not move around in box when moved. A foam insert, with slits cut into it for the electrodes to lay in, works well.
- Place CFEs in freezer (−20 °C frost free) to preserve life. Always store reference electrodes in a saturated (3 M) NaCl solution.

Phase 4: Cleaning and Nafion coating

- Fill one small (5 ml) beaker with isopropanol (2-propanol) and another with double-distilled (dd) water (H_2O).
- Submerge tip of CFE into the isopropanol (2-propanol) and swirl about ten times, then dip in dd H_2O ten times.
- Place on baking tray in oven and bake for 5 min at 200 °C. When placing the electrodes on tray, be sure to bend the wire such that the electrode tip is in the air and not touching the tray or other CFEs.
- Apply Nafion (5% solution, Aldrich) coating by dipping tip (only the tip) of CFEs into bottle of Nafion and swirling in a circular motion 10–12 times. Bake for 3 min at 200 °C. Repeat this step two more times (for a total of three coats of Nafion).
- Once cooled, place the CFEs in a storage container (again, plastic boxes containing a foam insert with slits cut into it for the electrodes to lay in works well). Store CFEs in freezer

(−20 °C frost free) until ready for calibration and use (otherwise Nafion coat will lose integrity). Coated CFEs are good kept in freezer for ~2 months.

2.4 Electrode Calibration

Prior to use in vivo, the CFE must be calibrated such that known concentrations of neurotransmitter correspond to the current generated by their oxidation at the surface of the CFE. In addition, the calibration process is important for assuring the effectiveness of the Nafion coat, which should prevent anionic species from coming into contact with the CFE. Our criterion is that each CFE must have at least 500-fold greater selectivity for the parent biogenic amine (5-HT, DA, NE) than for its corresponding metabolites (5-hydroxyindoleacetic acid (5-HIAA), 3,4-dihydroxyphenylacetic acid (DOPAC) and homovanillic acid (HVA), respectively) and must have at least 500-fold greater selectivity for the biogenic amines than for ascorbic acid. To determine this, and to calibrate, each CFE is placed in a beaker containing either PBS or aCSF, depending on what vehicle will be used in experiments. Once the current response to stepping the CFE potential stabilizes (within a few min), a possible interferent (i.e. anionic species ascorbic acid, 5-HIAA, DOPAC, or HVA) is added to a final concentration of 250 μM, followed by increasing concentrations of 5-HT, DA, or NE (typically 0.5 μM increments, ranging from 0.0 to 5.0 μM). In addition to having a biogenic amine to interferent selectivity ratio (e.g. 5-HT:5-HIAA) greater than 500:1, electrodes used for our in vivo experiments must have a linear response to graded concentrations of the neurotransmitter of interest ($r^2 > 0.90$); *see* **Note 3**. We typically find electrodes prepared in this way have a detection limit in the range of 30–60 nM.

The steps for calibrating CFEs follow:

First you will need to prepare buffer and solutions relevant for the neurotransmitter of interest. These include:

- PBS (0.1 M) (or aCSF; we find little difference between PBS and aCSF in regard to the calibration factors obtained, so most commonly use PBS for simplicity). For ingredients, see Sect. 2.
- Interferent (20 mM; 5-HIAA, DOPAC, HVA, or ascorbic acid). A volume of 5 ml is ample for calibration purposes.
- Analyte (1 mM; DA, 5-HT, or NE). Again 5 ml is more than ample.

Note that concentrated stock solutions of interferent (e.g. 0.1 M) and analyte (e.g. 20 mM) can be made and stored in Eppendorf tubes (1.7 ml) and frozen (−20 °C) until use. Stock solutions will last 5–6 months if kept this way. Also note that all solutions are made with 100 μM ascorbic acid as an antioxidant and pH adjusted to 7.4.

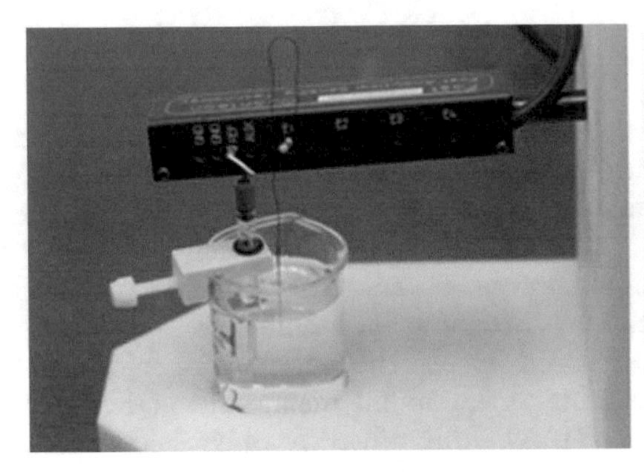

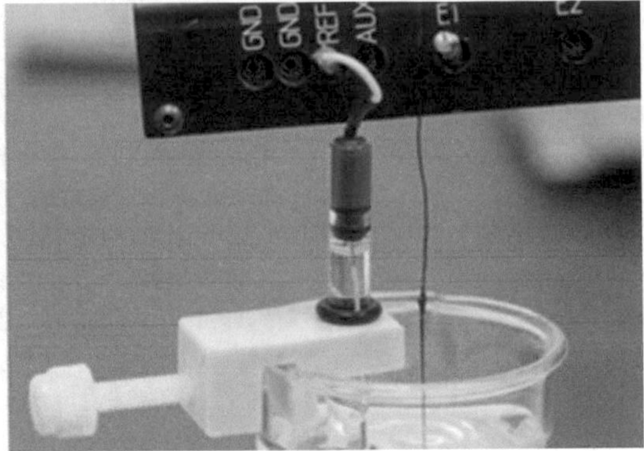

Fig. 1 Calibration set-up. We suggest having a custom-made stand (like the one pictured in *left panel*) to support the headstage and beaker during calibration. A clamp to hold the AgCl reference electrode can be fabricated or purchased from Thomas Scientific, Swedesboro, NJ, USA. Note that both the reference electrode and CFE are positioned to one side of the beaker (*right panel*). This is to allow room for mixing the solution between addition of interferent and successive additions of neurotransmitter of interest. Be very careful not to damage either electrode when mixing, and avoid formation of a vortex or bubbles

2.4.1 Procedure

- Add 40 ml of PBS to a 50 ml beaker.
- Lower AgCl reference electrode into beaker and connect to headstage (see Fig. 1).
- Lower CFE into beaker and connect to headstage (see Fig. 1).
- Follow the software instructions to begin calibration.
- Voltage (+0.55 V) is applied to the electrode (according to square wave described in Sect. 2.3 below). Allow recorded current to stabilize in PBS/aCSF solution for 15–20 min before beginning calibration. The current response should be very stable by this time.
- Begin calibration. Mark baseline signal (event mark) according to the software you are using. For FAST software from Quanteon, this is typically the computer's "HOME" key.
- Add 500 μl of interferent (20 mM) to the beaker and mix thoroughly (10–12 times), then make an event mark again. Your solution will now contain 250 μM of interferent. We find that using a Betty Crocker hand blender, with run down batteries, works well for mixing. Be careful when mixing not to break either CFE or reference electrode, nor to create a vortex. If you find that the recorded current response increases following addition of interferent, you will know that your Nafion coating has failed and the interferent is making contact with the CFE. In this case, dry the CFE thoroughly (in oven at 200 °C) and re coat with Nafion. If the Nafion coat fails again, discard the CFE.
- Add neurotransmitter in volumes that will ensure the CFE is calibrated in the relevant range. For example, if you expect that

the current recorded in vivo will correspond to a signal of ~1 μM peak amplitude, calibrate the electrode from 0 to 3.0 μM in 0.5 μM increments (i.e. six additions of 20 μl of 1 mM neurotransmitter stock solution, mixing the solution thoroughly between each addition). Of course, when starting out, the calibration range will need to be determined empirically. Note that speed is key to performing an optimal calibration. Do not wait longer than 10–20 s between consecutive additions of neurotransmitter. Because of the need for speed and because the volumes of neurotransmitter added are small relative to the 40 ml of PBS (+0.5 ml interferent), we do not increase the volume of neurotransmitter added upon sequential additions during calibration. In any case, the stepwise calibration error is negligible (i.e. <1.25 %).

- At the conclusion of calibration, rinse the electrode thoroughly with ddH_2O. Use tape, attached to the copper wire near the gold pin connector, to label the electrode for later use. Store as previously described until ready to use.

2.5 Measuring Clearance of Exogenously Applied Neurotransmitter *In Vivo*

One of the greatest advantages of locally applying neurotransmitter to brain is that the source and identity of the neurotransmitter is known. This is especially useful when the desired measure is that of transporter function, specifically in terms of its ability to *remove* neurotransmitter from the extracellular milieu without an associated release component. In this regard, measuring clearance of exogenously applied neurotransmitter provides a straightforward and readily interpretable approach. As described in the procedures below, this approach also allows the experimenter to vary the amount of neurotransmitter delivered to the extracellular fluid, thereby enabling calculation of maximal velocity (V_{max}) for neurotransmitter clearance and affinity values, in vivo, denoted here as K_T (see also, Sect. 2.6). Given that synaptic levels of neurotransmitter are estimated to be in the millimolar range and extrasynaptic levels in the nanomolar to micromolar range [10–12], CFEs are capable of detecting physiologically relevant concentrations of neurotransmitter.

In vivo chronoamperometric recordings typically take a full 8 h day (or longer, depending on your experiment and how well you can maintain your animal under anesthesia). There are numerous steps, which should be followed in sequence.

2.5.1 Step One

Chloride plating of reference electrodes for use in vivo.

You will need:

- Plating bath, consisting of 1 M HCl (prepared in distilled water) and saturated with NaCl.
- A 0.5–2.0 A 12 V power supply.
- Two Teflon-coated silver wires (0.008″ bare).

Start by stripping 4–5 cm of Teflon from one wire and about 1 cm from the other wire. These wires should have the same gold pin connectors that are used on the carbon fiber working electrodes.

- Connect the wire to be plated (with 1 cm bare) to the anode (+) of the 12 V power supply.
- Connect the wire with the longer (4–5 cm) bare end to the cathode (−) of the 12 V power supply.
- Place both wires in the HCl/NaCl bath and turn power on. Watch for bubbles to roll off counter electrode.
- Allow plating to proceed for 15–30 min (or longer).
- Place the chloridized tip of the newly plated reference electrode in a solution of 3 M NaCl and compare against a stable reference.
- Store reference electrodes in 3 M NaCl until ready to position in your preparation.

2.5.2 Step Two

Prepare solutions

- Vehicle (typically PBS, but also aCSF; see Sect. 2 for recipes).
- Neurotransmitter (e.g. 5-HT, DA, NE) solutions. We find that filling the glass micropipette barrels with a 200 μM solution of neurotransmitter is a good starting concentration. The tips of the multi-barreled micropipette and CFE are separated by 200–300 μm (see step 4). We find that when neurotransmitter is pressure-ejected into brain, the concentration of neurotransmitter reaching the CFE is reduced by ~200-fold; thus for a barrel concentration of 200 μm, the current produced by oxidation of neurotransmitter at the tip of the CFE will produce a signal equivalent to ~1 μM.
- Drugs. These will be whatever your experiment demands. For example, if examining the effect of a selective serotonin reuptake inhibitor (SSRI) on 5-HT clearance you would fill one (or more barrels) of the multi-barreled pipette with the SSRI of your choice (*see* **Note 4**). We find that 400 μM is a good starting barrel concentration. This can be adjusted empirically according to the design of the experiment and results as they come to hand (*see* **Note 5**).
- Be sure to add ascorbic acid (100 μM) to solutions as an antioxidant.

2.5.3 Step Three

Fill multi-barreled micropipette

- Remove multi-barreled micropipette from Plexiglas box (Frederick Haer Co., FHC; www.fh-co.com); four barrels (FHC order specifications: 14-15-L; tip inner diameter modification = 9–12 μm); seven barrels (FHC order specifications: 17-15-L; tip inner diameter modification = 9–12 μm).

- Number or color-code each barrel of the micropipette with a Sharpie for identification purposes.
- Fill each barrel of the micropipette with the desired neurotransmitter/drug/vehicle combination using a 1 cc syringe, fitted with a "Microfil" tip to fill the pipette (MF34G-5, 34 gauge; 67 mm, World Precision Instruments, www.wpiinc.com). We find that securing the multi-barreled micropipette to the edge of a counter top with plasticine clay works well for filling purposes. When filling each barrel be sure to avoid trapping air bubbles in the barrel (this will make it impossible to accurately determine the volume of fluid you are pressure-ejecting into brain). Also avoid overflow. You do not want adjacent barrels contaminated with the contents of others.
- Once filled, use glue (Locite glue gel; Henkel Corp., Westlake, OH, USA) to adhere polyethylene tubing (PE10 end, see following) to each barrel of the micropipette. Allow glue to dry for at least 10 min. The length of tubing connecting the micropipette to the picospritzer consists of two different diameter lengths of tubing, PE10 that is connected to PE50, and held together with glue gel. The length of PE50 tubing needs to be sufficient to comfortably reach and connect to the picopritzer's pressure port. PE10 tubing, which is the end glued into the micropipette, can start being as long as 30 cm. In this way you can simply cut the tubing near the base of the micropipette at the end of each experiment and reuse the same lengths of combined PE10, PE50 tubing for many experiments.
- Label tubing (PE50 end) using tape so that the content of each barrel can be easily identified. Again, note that PE10 tubing is inserted into the glass micropipette barrels and glued; the PE50 ends will be coupled individually to the picospritzer pressure port when delivery is desired (tubing is shown in Fig. 2).

2.5.4 Step Four

Making the multi-barreled micropipette/CFE assembly

- Place filled multi-barreled micropipette in one arm of micromanipulator and the CFE in the other (see Fig. 2). Using a dissecting microscope, fitted with an eye piece micrometer to view the CFE and micropipette tips, adjust so that the tips are 200 or 300 μm apart (for mouse and rat, respectively). Ensure that the tip of the multi-barreled micropipette and CFE are aligned and in the same plane. It is critical for tip separation and alignment to be consistent among experiments as differing distances and dimensional planes between the tips of each will greatly influence the time it takes for ejected compounds to reach the CFE and their concentration measured at the CFE surface—both are important factors for proper interpretation of data.

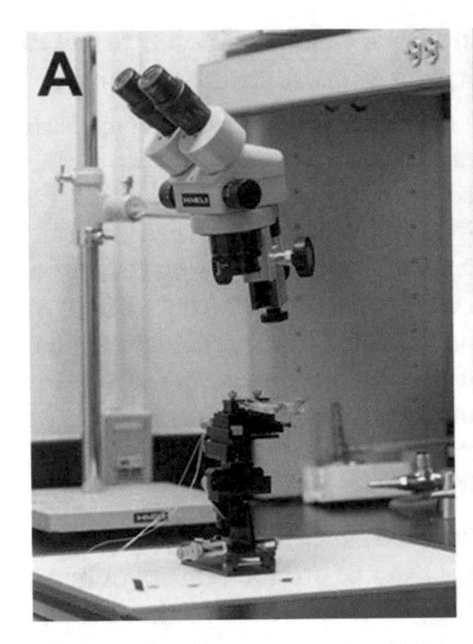

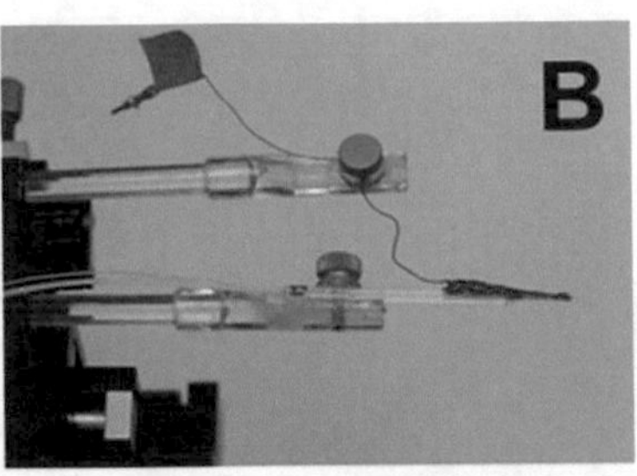

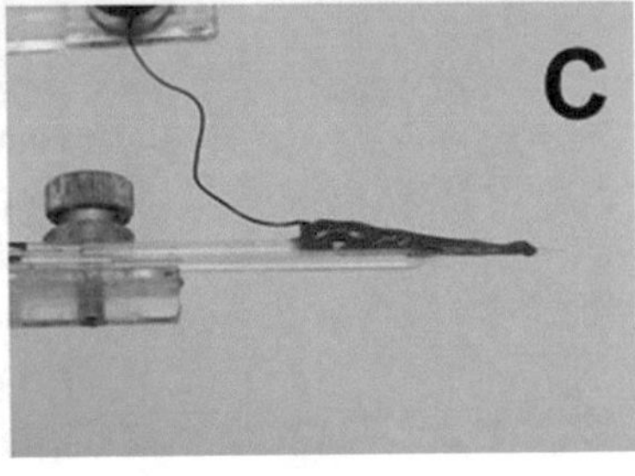

Fig. 2 CFE-multi-barreled micropipette assembly. Secure the CFE into one arm of micromanipulator and the multi-barreled micropipette into the other. Using the micromanipulator, and looking through a dissecting microscope, position the tips of the electrode and micropipette such that they are either 200 μm (for mouse) or 300 μm (for rat) apart and aligned in the same plane. Secure the assembly with sticky wax. (**a**) Micromanipulator holding CFE and multi-barreled micropipette (with PE tubing attached; note the tape labeling the PE tubing laying on benchtop) and microscope to determine tip separation. (**b**, **c**) Show configuration of micropipette assembly, fastened with sticky wax. Note the multi-barreled micropipette labeled with Sharpie. In panel **c**, the menisci in the multi-barreled micropipette can also be seen

- Turn on Bunsen burner and melt sticky wax (Kerr; Dental Sticky Wax). Secure electrode to pipette by gently letting molten sticky wax drip onto assembly (see Fig. 2). Do not wax closer than 8 mm from tips (this will impede lowering assembly to deeper brain structures). Wax both sides of the assembly. Allow 5–10 min for curing.

2.5.5 Step Five

Preparing the animal and recording

- Prepare anesthesia. We frequently use a cocktail containing urethane (1.0 g/ml) and α-chloralose (0.1 g/ml) at a dose of 850 mg/kg urethane and 85 mg/kg for rats, and 350 mg/kg and 35 mg/kg for mice. To make, combine 1 g urethane with 0.1 ml normal saline and warm to ~40 °C, then add chloralose and vortex/stir until dissolved. This anesthesia is long-acting (so supplementing anesthesia during the experiment is rarely required) and does not interfere with biogenic amine transporters. Anesthesia can be prepared fresh, or aliquots can be frozen in Eppendorf tubes and thawed before use. Frozen anesthetic is good for ~2 months. *Important note*: Prepare anesthetic under a fume hood, wearing gloves, and a face mask as a precaution. Urethane is carcinogenic and fumes should not be inhaled.

- Prior to retrieving the animal from the animal facility, ensure heating pads are turned on and warm. This includes the water-circulated heating pad, which will maintain the animal's body temperature during the experiment, and an electric heating pad, which will keep the animal warm while the anesthesia is taking effect.
- Once anesthesia has been administered intraperitoneally, monitor the animal carefully. Check for appropriate level of anesthesia by checking for toe or tail pinch reflex in mice, and for the foot pinch or corneal reflex in rats.
- Once the animal is fully anesthetized, insert a tracheal tube. This facilitates the animal's respiration during the course of the experiment. Place the animal supine and gently tape forelegs and head to bench surface. Make a ventral longitudinal incision (~1 cm) in the skin of the neck, being careful not to sever major blood vessels. Using fine tipped forceps, tease away the musculature to expose the trachea. Using Vannus scissors, make a small cut in the trachea (between the third and fourth tracheal cartilage rings, inferior to larynx). Do not cut the trachea all the way through. Posterior to the cut, run two, ~8 cm long lengths of suture (5-0 silk) under the trachea. Insert stainless steel tube into trachea and secure with suture. Trim excess suture and close incision with wound clips.
- Place animal in a stereotaxic frame modified for rat or mouse as appropriate (Stoelting; David Kopf).
- Using a #10 scalpel, make an incision from rostral to caudal the entire length of the sagittal suture on skull. With gauze, reflect the galea apponeurotica and associated connective tissue laterally until the skull is clearly visible. Use a thermocautery tool to stop any bleeding.
- Ensure the head is properly secured and aligned in the stereotaxic frame before proceeding. This should be done by ensuring that the skull is leveled between bregma (intersection of coronal and mid-sagittal sutures on skull) and lambda (lambdoidal suture) landmarks. We highly recommend that a previously used CFE-micropipette assembly is used to determine bregma and lambda, to reduce risk of damaging the assembly that will be used for the experiment. It is helpful to use a stereoscope, mounted on an adjustable boom arm, to visualize the tips of the CFE-micropipette assembly and reduce the risk of breaking them on the skull.

- Exchange the used CFE-micropipette assembly in the arm of the stereotaxic frame with the newly prepared assembly to be used for the experiment. Note that stereotaxic coordinates for bregma for your experiment should be determined using this assembly. *Be careful not to touch the tip of the assembly to the skull, this can damage either the Nafion-coated carbon fiber or the glass tip of the micropipette.* Using a stereotaxic atlas (e.g. 13), set the stereotaxic arm at the coordinates required to target your brain region of interest. Lower assembly to within a few mm of the skull's surface and mark area of skull that will be removed using a pencil.
- Raise assembly and swing stereotaxic arm out of the way while performing a craniotomy to remove skull over brain region of interest. Use a dremel tool for the craniotomy being careful to drill gently so as not to push through the skull and into underlying brain. This is especially important if using mice or young animals (rat or mice) where the cranium is much thinner and softer. Using fine forceps or dura hook, remove meninges.
- With the brain area of interest now exposed, swing the stereotaxic arm holding the CFE-micropipette assembly back into place and lower to the relevant dorsal-ventral coordinate using a microdrive (Narishige; MO-10). Be careful to lower the assembly very slowly. If the assembly is lowered too quickly, the distance between the tip of the CFE and multi-barreled micropipette can increase or shift out of plane. This can lead to an inability to detect any exogenously applied neurotransmitter at the recording site of interest, which then requires removal of the assembly for reconfiguration.
- Once the CFE-micropipette assembly is in place, insert the tip of the silver wire reference electrode under the superficial scalp contralateral to the recording electrode.
- Apply square wave voltage steps according to neurotransmitter being studied. For 5-HT (see Fig. 3), the CFE is held at a zero resting potential vs. a Ag/AgCl reference electrode. Then the potential is stepped to +0.55 V. This voltage step is selected to exceed the peak 5-HT oxidation potential by 0.15 V. The square wave voltage step is held at +0.55 V for 100 ms and this produces a rapid increase of current (typically in a range of 1–3 nA) at the CFE, the amplitude of which is proportional to the number of electrons removed from neurotransmitter molecules (i.e. oxidation) as a function of time during the applied voltage step (~4 electrons per 5-HT molecule). Measured current decays as a function of time, as neurotransmitter is oxidized and/or diffuses away from the CFE. Note that the initial spike of current reflects charging of the electrode capacitance and is permitted to decay (typically 20 ms) based on the CFE time constant before the oxidation current is sampled and

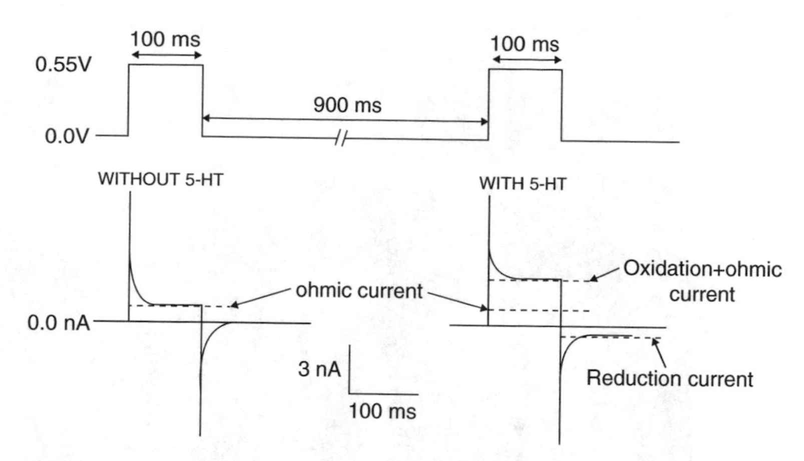

Fig. 3 Square wave voltage pulses applied to CFE for high-speed chronoamperometric recording of 5-HT. Ohmic current recorded without pressure-ejection of exogenous 5-HT (*left*) is set as "zero" and the change in current produced by oxidation of exogenous 5-HT (*right*) at the electrode is measured as a function of time. The current produced 20–100 ms after the potential is stepped to +0.55 V and then again after it is stepped back to 0.0 V is digitally averaged

averaged for 80 ms. The square wave voltage step is then returned to 0.0 V and held at this potential for 900 ms. This return to the resting potential gives rise to a reduction current, which for 5-HT and its oxidation products is close to zero. For other neurotransmitters such as dopamine, the applied and resting potentials are typically each 100 ms in duration. However, unlike dopamine, which is readily oxidized and reduced, serotonin is "reluctantly" reduced. One important implication of this is that the lack of sizable "reverse" current can cause "fouling" of the CFE (*see* **Note 3**).

- Allow ~10–20 min for the current to stabilize.
- Begin experiment according to relevant design (see example below and see Fig. 4 for annotated photograph of experimental setup).
- Throughout, keep exposed brain moist with saline.

2.5.6 Example Experimental Paradigm

Important note: The amount of neurotransmitter or drug delivered can be quantified by measuring the amount of fluid displaced from the micropipette using a dissection microscope fitted with an eyepiece reticule (Fig. 4). This can be simply calculated using the equation $V = \pi r^2 \times l$, where V is the volume displaced, r is the inside radius of the pipette barrel, and l is the distance the meniscus falls within the barrel during pressure application. Given that the concentration of each solution contained within the barrel is known, and the volume ejected is readily measured, the moles of neurotransmitter or drug delivered can be easily calculated.

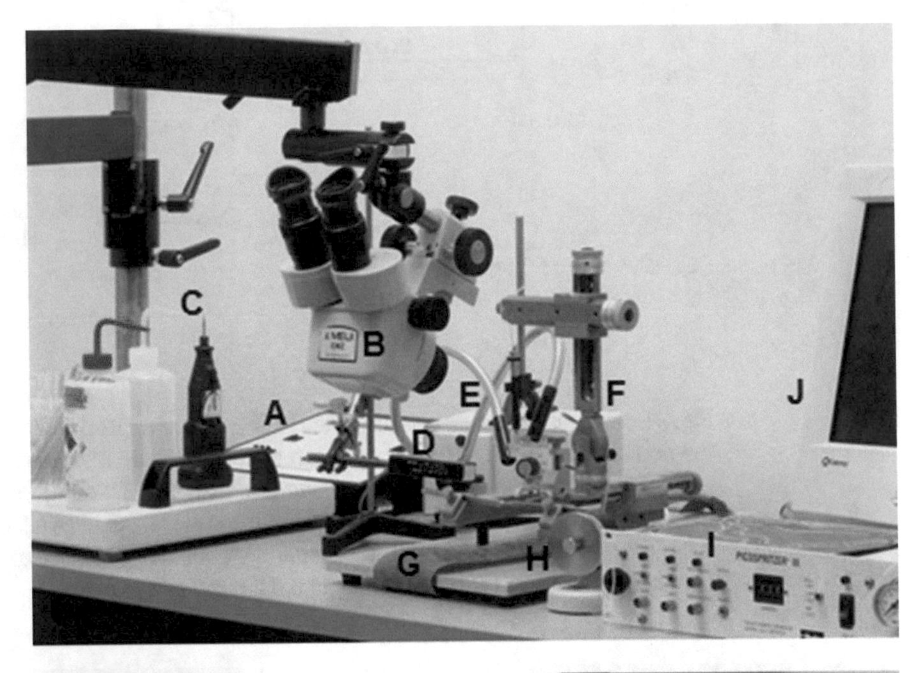

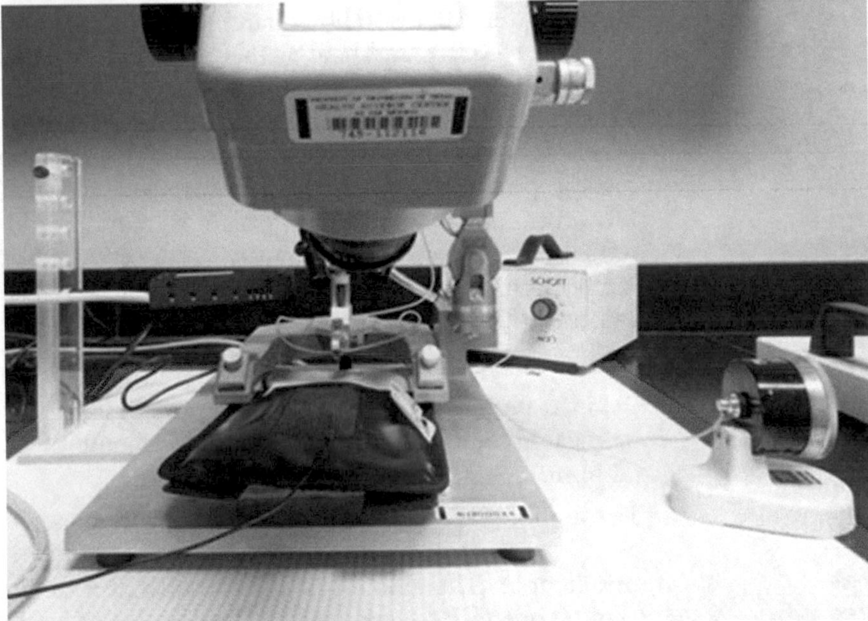

Fig. 4 Experimental set-up. *Upper panel* showing, *A* potentiostat, *B* dissecting microscope on boom stand, *C* dremel tool, *D* headstage, *E* light source, *F* stereotaxic frame, *G* water-circulated heating pad, *H* microdrive, *I* picospritzer, *J* monitor (and computer, out of picture). *Lower panel*, showing a closer view of headstage, stereotaxic frame, water-circulated heating pad, microdrive, and dissecting microscope, which is used both during surgical preparation of the animal, and for measuring the fall of meniscus in the micropipette

Experiment: *Measuring the effect of a drug* (*see* **Note 6**) *on clearance of exogenously applied 5-HT*. In this scenario, it is best to use a seven-barreled micropipette (though a four-barreled pipette may also be used). The key advantage to using seven barrels is that different concentrations of drug can be loaded into five of the seven barrels, with 5-HT and vehicle loaded into the remaining two barrels. In this way, different amounts (typically pmol amounts) of drug can be delivered in equivalent volumes by pressure-ejection. This eliminates any potential confound that might result from pressure-ejection of different volumes of the same barrel concentration of drug to deliver a given pmol amount. It also reduces variability resulting from local heterogeneity of SERT expression. That said, it is also wise to consider filling two barrels with 5-HT, two with vehicle and the remaining three with the same, or differing concentrations of drug (e.g. for dose–response analysis). In this way, there is a "back-up" 5-HT and vehicle barrel should one barrel become blocked (*see* **Note 7**).

The first step is to establish a reproducible signal. Pressure-eject 5-HT into the brain region of interest (e.g. CA3 region of hippocampus). Typical pressure settings range from 5 to 25 psi, for a duration of 0.25–3 s. The settings for any given experiment will need to be determined empirically and adjusted to deliver the pmol amount of 5-HT needed to achieve a signal with the desired peak amplitude. Signals with amplitudes in the range of 0.25–0.5 μM are most likely to engage active uptake by SERT, whereas concentrations upward of 1.0 μM will begin to recruit activity of other transporters (including NET and organic cation transporters (OCTs)), at least in hippocampus. If primary interest is in studying the high-affinity transporter for a given neurotransmitter, then it is best to keep peak signal amplitudes as small as can be reliably detected.

We allow a minimum of 3 min to elapse between ejections, but most often keep the inter-ejection interval at 5 min. This avoids sequential application of exogenous agonist influencing the activity state or plasma membrane expression of the transporter being studied (see 14, 15). We routinely find that 3–5 ejections of serotonin into hippocampus are needed before the signal becomes reproducible. Before this stable baseline is achieved, the signal amplitude for the same pmol amount of 5-HT declines and signal time course decreases modestly. Part of this is no doubt attributable to initial disruption of the extracellular matrix by the bolus ejection of 5-HT that could change factors known to increase diffusion (e.g. decreased tortuosity, increased volume fraction) and hence accelerate disappearance of 5-HT from the vicinity of the CFE. In addition, we believe that trafficking of SERT to the plasma membrane in response to the rapid increase in extracellular 5-HT also contributes. This phenomenon has been very well characterized by Randy Blakely and his group using in vitro models (e.g. 16–18). We find that once we have attained stability in the signal produced by repeated pressure-ejection of 5-HT into hippocampus, the signal remains remarkably stable for very long periods (hours). This provides an excellent preparation for studying the ability of drugs to impact various signal parameters.

At this point, the drug of interest is now pressure-ejected into brain. We typically allow 1–2 min to elapse, to ensure the drug has fully diffused to the area surrounding the recording electrode, before pressure-ejecting the same pmol amount of 5-HT. At this point three things can happen, (1) nothing (i.e. the drug has no effect), in this scenario wait 5 min or longer and then apply 5-HT again (sometimes drugs have a delayed effect). If still no change in signal and a set of three replicate signals are obtained, move onto the next dose, or the next drug (or vehicle) to be tested; (2) clearance time increases or (3) decreases (with potentially corresponding decreases and increases in peak signal amplitude, respectively). In this scenario, continue pressure-ejecting 5-HT at 5 min intervals (or longer, if clearance time is increased), until the 5-HT signal parameters return to their pre-drug baselines. In this way you will have both a time course of drug action, and be able to establish a new set of reproducible signals so as to set baseline before testing the next dose, or drug. An important precaution for constructing accurate dose–response curves is to deliver the full range of doses in a randomized sequence to avoid the possibility of desensitization tachyphylaxis. Continue in this way until all drugs and doses for that experiment have been tested. Note that collecting all data points (e.g. doses for full dose–response curve; testing of multiple drugs) is uncommon within a single animal (*see* **Note 8**).

Once data collection is complete, mark placement of the CFE by making an electrolytic lesion (for procedure, see Sect. 2.5), and remove brain for histological verification of electrode placement (see Sect. 2.5). Be sure to save all data as well as backup files on an external hard drive (or alternative).

There are many examples of this application of chronoamperometry including [19–30] and more.

2.6 Measuring Release and Clearance of Endogenous Neurotransmitter In Vivo

High-speed chronoamperometry can also be used to measure the kinetics of stimulated release and clearance of endogenous neurotransmitter. This can be achieved in a variety of ways. For example, by pressure-ejecting a depolarizing agent such as potassium chloride, or a drug that causes neurotransmitter release such as amphetamine, or by electrical stimulation of synaptic inputs. In these scenarios, the nature of neurotransmitter being measured is not as well-defined as when a specific neurotransmitter is exogenously applied. Evoking release of endogenous neurotransmitter can result in a "soup" of neurotransmitters being released. This can be overcome by setting the experimental paradigm to favor measuring release of a particular neurotransmitter, for example, measuring amphetamine evoked neurotransmitter release in dorsal striatum to favor the majority of evoked release being DA. Regardless, close attention needs to be paid to the chronoamperometry "signature" of the neurotransmitter. The signature is the ratio of reduction and oxidation (red:ox) currents. For 5-HT this is typically less than 0.1 (due to the lack of reverse, or reducing, current). For NE this is typically 0.1–0.3 and for DA, typically 0.6–1.0. While red:ox currents help to provide initial identification of the neurotransmitter released, there does exist a "grey zone". For example, studies by Hoffman et al. [31], measuring stimulated release of neurotransmitter in substantia nigra, show that red:ox ratios of ~0.5 comprise significant amounts of 5-HT and so considered only signals with red:ox values of ≥1.0 to be DA. In sum, if using chronoamperometry to measure evoked-release of neurotransmitter it is critical to pay attention to the red:ox "signature" when interpreting data, and to pharmacologically verify the neurotransmitter(s) contributing to the signal, if need be.

2.6.1 Procedure

Steps one through five are as for Sect. 2.3 above, with the obvious exception that barrels of the micropipette will now be filled with a neurotransmitter releasing agent (e.g. KCl, amphetamine), and vehicle.

- Vehicle (PBS or aCSF).
- KCl (70 mM KCl, 79 mM NaCl, 2.5 mM CaCl; pH—7.2–7.40).
- AMPH (400 μM made in aCSF).
- Any releasing compound of interest, or electrical stimulation (not discussed here).

Example Experimental Paradigm: *Measuring the effect of a kinase inhibitor on amphetamine-evoked DA release in dorsal striatum.* First establish a reproducible baseline amphetamine-evoked DA release. The volume pressure-ejected will need to be determined empirically to yield signals of the desired peak signal amplitude. Be sure to check the red:ox values. This value should be ~0.6 or more for DA and should correspond with the red:ox values obtained when calibrating the electrode to DA. We find that 45 min intervals between applications of amphetamine are required to achieve reproducible signals. If the interval is shorter, DA stores become depleted and subsequent amphetamine-evoked DA release events become smaller. Once a reproducible baseline is established, pressure-eject kinase inhibitor, allow 1–2 min to elapse, then pressure-eject amphetamine again, and at 45 min intervals until the DA release produced by amphetamine returns to the pre-kinase inhibitor baseline.

There are many examples of this and similar applications of chronoamperometry [32–38].

2.7 Concluding In Vivo Recording and Histological Verification of Electrode Placement

2.7.1 Marking Electrode Placement

Once your experiment is complete, you will need to make an electrolytic lesion to mark the recording site. The tip of the CFE is so small that it is difficult to visualize histologically where precisely the electrode track ends, hence the need to lesion the site. For this you will need the following materials:

- 9 V battery.
- Two lengths (~15 cm) of PVC-coated wire (105 °C temp/300 *V*max, Consolidated Electronic Wire and Cable, Franklin Park, IL, USA) with ~1.5 cm from each end stripped.
- Alligator clips.
- PBS (0.1 M).
- Gauze pad.
- Timer.

Remove reference electrode from the animal.

- Unplug the CFE from headstage (do *not* move the CFE from its location in brain).
- Using alligator clips, connect one end of the PVC-coated wire to the positive lead of 9 V battery and the other end to the gold pin connector of the CFE.
- Wrap animal's tail with gauze that has been thoroughly dampened with PBS.
- Using alligator clips, connect one end of PVC-coated wire to the negative lead of 9 V battery and other end of wire to the gauze surrounding animal's tail.
- Leave for 30 s, then remove wires and CFE electrode from brain.

2.7.2 Storing the Brain for Histological Verification

Next the brain will need to be stored for later histological confirmation of CFE placement. To remove the brain and prepare it for storage you will need the following:

- Ice bucket.
- Ice.
- Petri dish.
- Physiological saline.
- 50 ml beaker.
- Fine forceps.

- Scissors.
- Filter paper.
- Small animal guillotine.

Place ice in bucket and chill a beaker filled with saline.

- Chill forceps, scissors, and lid of petri dish on ice.
- On ice, place a sheet of filter paper on lid of petri dish and dampen with saline.
- Decapitate animal.
- Remove brain and place in beaker of chilled saline for 2–5 min.
- Place brain on filter paper to remove meninges with forceps.
- Place brain in powdered dry ice until frozen.
- Place frozen brain in labeled cryo-bag and store in –80 °C freezer (at least overnight).

2.7.3 Histological Verification

Finally, you will need to section the brain and stain it for histological verification of electrode placement. For this you will need the following:

- Cryostat (e.g. Leica, Model CM1850 cryostat, Buffalo, Grove, IL, USA).
- Gelatin-coated slides (chilled overnight in –20 °C freezer) and slide box.
- Desiccator with vacuum line.
- Thionin stain (see recipe below).

Remove brain from –80 °C freezer and place in cryostat (with temperature set between –18 and –20 °C), together with gelatin-coated slides, for 1 h.

- Slice brain in 20 μm coronal sections.
- Heat fix brain section to slide using gloved hand (by rolling back of slide over finger or hand slowly and gently to reduce the risk of air bubbles being trapped under the brain section).
- Place in slide box and store overnight in desiccator (with vacuum on).
- Remove slides from desiccator, stain with thionin and coverslip (see below).
- View under microscope to determine electrode placement.

Thionin Staining

Acetic Acid (1 M) (at least 100 ml stock solution).

- NaOH (1 M) (at least 100 ml stock solution).
- ddH_2O.
- 500 ml beaker.

- Thionin (Fisher Scientific).
- Hot plate with stirrer.
- Stir bar.
- Vacuum filter.
- Filter paper.

To Make Thionin Stain

Add 50 ml 1 M acetic acid and 19 ml 1 M NaOH to beaker.

- Place on hot plate and stir (with stir bar).
- Bring to 250 ml with ddH_2O.
- Adjust to pH 4.6.
- Heat until steaming.
- Add 2.5 g thionin (while stirring).
- Stir on low setting for 1 h.
- Vacuum filter.
- Store at room temperature.

To Stain Brain Sections You Will Need the Following

Staining dishes × 14.

- Slide racks.
- Flat tipped forceps.
- 100% EtOH.
- 95% EtOH.
- 70% EtOH.
- 50% EtOH.
- ddH_2O.
- Citra-sol (Hemo-De).
- Resin for mounting (Permount toluene mounting media; Fisher Scientific), coverslips, and Pasteur pipette.

Procedure

Fill five trays with 100% EtOH, two trays with 95% EtOH, two trays with 70% EtOH, one tray with 50% EtOH, two trays with Hemo-De, one tray with ddH_2O, and one tray with thionin stain.

- Load slide racks with slides containing brain sections.
- Stain according to the following steps:

– 100% EtOH	2 min
– Hemo-De	3 min
– 100% EtOH	1 min
– 100% EtOH	1 min
– 95% EtOH	2 min

– 70% EtOH	2 min
– ddH_2O	2 min
– Thionin	5–30 s
– 50% EtOH	Dip
– 70% EtOH	Dip
– 95% EtOH	Dip
– 100% EtOH	Dip
– 100% EtOH	2 min
– Hemo-De	2–3 min

- Dry momentarily (~30 s).
- Coverslip using resin and allow 7 days to cure.

2.8 Data Analysis

Currents (converted to micromolar concentration based on the calibration factor (slope of linear regression) calculated for each CFE) can be analyzed in several ways using a variety of components of the signal. Some of the more frequently analyzed signal parameters are shown in Fig. 5 and include the peak signal amplitude, rise time (the time for the signal to reach its peak amplitude), the time for the signal to decay by 20, 50, 60, and 80% of its peak amplitude (T_{20}, T_{50}, T_{60}, and T_{80}), area under the curve, and clearance rate.

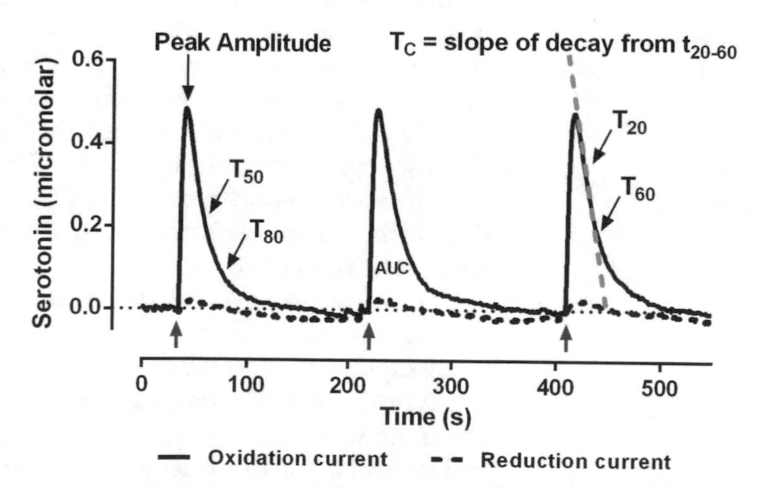

Fig. 5 Signal parameters and reproducibility of electrochemical signals produced by local application of 5-HT in the CA3 region of the anesthetized mouse. Serotonin (2 pmol, indicated by *arrows* along the lower abscissa) was applied at ~3 min intervals by pressure-ejection. Shown is the oxidation current and reduction current converted to micromolar concentration using calibration factor determined in vitro. Signal parameters indicated include peak signal amplitude, T_{20}, T_{50}, T_{60}, T_{80}, T_c, and area under the curve (AUC)

We and others (e.g. 22, 36, 39) calculate clearance rate in two ways. One method is by calculating the slope of the most linear portion of descending phase of the signal, Tc. For 5-HT, DA, and NE signals, this most often occurs between T_{20} and T_{60}. The other is by calculating the velocity, V, for neurotransmitter (NT) clearance by fitting the entire descending portion of the signal to an equation describing one-phase exponential decay,

$$Y = [\mathrm{NT}]_{\mathrm{o,i}} \cdot \mathrm{e}^{-k \cdot t} + [\mathrm{NT}]_{\mathrm{o,t}}$$

where,

Y is the concentration of NT (μM)

$[\mathrm{NT}]_{\mathrm{o,i}}$ is the initial maximal extracellular concentration of NT (μM)

$[\mathrm{NT}]_{\mathrm{o,t}}$ is the extracellular concentration of NT (μM) at baseline

k is the rate constant (s^{-1}).

The half-life of decay is $0.6932/k$. Note that because the concentration of NT at baseline (i.e. $[\mathrm{NT}]_{\mathrm{o,t}}$) equals "zero" (the value given to represent the baseline electrochemical signal *prior* to addition of exogenous NT, or release of endogenous NT), this term of the equation reduces to zero and is not required for the analyses. The calculated k value from this equation is then multiplied by the peak amplitude of the signal to give the velocity of transport (V). These calculations can be made for signals of increasing amplitude until an apparent V_{max} is reached and K_T can be determined. Note that K_T is equivalent to K_m, however denoted differently to distinguish this as an affinity value calculated in vivo. K_T can be corrected for volume fraction (α), which in most regions of brain is 0.2 (see 22). Even correcting for α, K_T values are typically greater than K_m values determined in vitro due to the inherent contribution of diffusion to clearance profiles obtained in vivo.

Of course, more sophisticated analyses can also be performed; for example, fitting data to equations describing two- (or more) phase exponential decay. This type of analysis is helpful for parsing out contributions of more than one transporter in mediating neurotransmitter clearance (e.g. high-affinity, low-capacity transporters and low-affinity, high-capacity transporters) or more than one modulatory process grading the performance of a single transporter.

3 Notes

1. Advantages of the Quanteon system: It is "turn-key", meaning the system is ready to go, equipped with its own data

acquisition and analysis software and operating programs for chronoamperometry (as well as other forms of voltammetry). The staff at Quanteon are very helpful in troubleshooting issues with their product. Disadvantages: Occasionally problems cannot be resolved. A self-assembled system allows greater flexibility and power in what your system can ultimately do, but requires a sound knowledge of assembling and operating electrochemistry/electrophysiol ogy instrumentation, as well as interfacing them with appropriate software.

2. Pre-fabricated electrodes from Quanteon are relatively expensive in comparison to those fabricated as described in Sect. 2. Shipping can sometimes damage pre-fabricated electrodes so their functionality cannot be guaranteed. The materials to make CFEs are relatively inexpensive. Training an individual to fabricate Nafion-coated CFEs takes some time, but in the long-run, is cost-effective. Assess the cost–benefit for your needs.
3. Note that linearity of response to increasing concentrations of 5-HT and NE is excellent over the range of 0.0–3.0 μM. If calibrating to higher concentrations of these neurotransmitters (e.g. >5.0 μM), the calibration becomes curvilinear, likely due to adsorption of electrogenerated products to the electrode [40, 41]. Even so, regression analysis continues to yield strong r^2 values (~0.90) showing the utility of chronoamperometry to measure relatively high concentrations of 5-HT and NE. We find that if we limit exposure of the electrode to these higher concentrations of 5-HT or NE (1–2 exposures), we can effectively establish values for maximal rates of 5-HT or NE clearance (V_{max}) in various brain regions in vivo (e.g. see 22). If more than 1–2 exposures of high (>3.0 μM) 5-HT or NE are required to fulfill the objective of your experiment, then expect to replace electrodes due to fouling.
4. When selecting relevant drugs for your studies, be sure to consider solubility. Drugs with poor solubility will cause the tip of the micropipette to become blocked, necessitating replacement of the micropipette and reassembly of the CFE-micropipette assembly. Certain drugs may also require vehicle containing DMSO or ethanol, both can affect neurotransmitter clearance in sufficiently high concentrations. In this scenario, be sure to fill a barrel with DMSO- or ethanol-containing vehicle controls, just as you would with PBS or aCSF controls.
5. While measuring clearance of exogenously applied neurotransmitter has its advantages, it can be difficult to investigate the

effect of receptor agonists on neurotransmitter clearance. By virtue of pressure-ejecting neurotransmitter (agonist) into brain, native transporters and receptors for that neurotransmitter are likely activated. This makes detecting further activation by other agonists at those receptors difficult. In contrast, effects of receptor antagonists can be more easily observed. To enhance the ability to detect reliable effects of receptor agonists on neurotransmitter concentrations in vivo, we find it best to reduce the barrel concentration of neurotransmitter four-fold and eject volumes into brain that yield the smallest, reliable signal (~100–200 nM peak signal amplitude).

6. Prior to testing any drug in vivo, check that it is not electroactive over the range of voltages that you intend to apply in order to quantify the concentration of your test neurotransmitter. This can be done using the calibration set-up, but now add drug (to yield a concentration similar to that expected to reach the CFE in brain; e.g. ~2 to 40 μM for a barrel concentration of drug of 400 μM). It is also wise to perform a "concentration-effect" curve in vitro, to determine if the drug is detectably electroactive at only certain concentrations and if it "fouls" the electrode. This can be determined by calibrating the CFE to the neurotransmitter of interest before and again after the drug has been added to the beaker and comparing these calibrations to that for a CFE calibrated twice with PBS being added to the beaker in between calibrations.

7. While pressure-ejecting neurotransmitter or drug, it is important to monitor the fluid in the barrel being pressurized for ejection and from the other barrels as well. This will allow identification of any septal defect in the multi-barreled micropipette that allows exchange of fluid between or among barrels. For example, as the meniscus of the fluid being pressure-ejected goes down, you might occasionally see the meniscus in another barrel go up. This is clearly problematic and calls for removal and replacement of that multi-barreled micropipette. For this reason it is a good idea when filling a multi-barreled micropipette to stagger the menisci of each barrel, while keeping them close enough together that they can all be visualized within the same field of view through the microscope. For data from any experiment to be interpretable, the meniscus of only that barrel to which you are applying pressure should move. If your experiment calls for measuring effects of more than one type of drug, it is also wise to run a subset of experiments with only one type of drug loaded into the barrels. Results from such internal con-

trol experiments will firmly rule out (or in) the possibility of cross-drug contamination among barrels.

8. If the animal's body temperature and blood oxygen saturation levels are maintained, and the animal appears in overall good condition, recordings can last for up to 8 h. However, if the animal's body temperature or blood oxygen levels fall below those that can sustain normal transporter function (i.e. 36 °C and 90% arterial O_2 saturation), terminate the experiment. Data collected under such conditions will be difficult to interpret. In many cases, it may be desirable to also monitor the animal's blood pressure and heart rate during recordings to ensure they remain stable and within an acceptable physiological range (mean arterial pressure: 80–100 mmHg, heart rate: 500–600 beats per minute for mice).

Acknowledgments

Studies described herein were funded in part by NIH grants MH64489, MH093320, MH106978, DA18992, and DA014684 to L.C.D. and NIH grants HL102310 and HL088052 to G.M.T. The authors gratefully acknowledge Lester Rosebrock for photography.

References

1. Kissinger PT, Hart JB, Adams RN (1973) Voltammetry in brain tissue: a new neurophysiological measurement. Brain Res 55:209–213
2. Cottrell FG (1902) Z Physik Chem 42:385
3. Gerhardt GA, Oke AF, Nagy G et al (1984) Nafion-coated electrodes with high selectivity for CNS electrochemistry. Brain Res 290:390–394
4. Gerhardt GA, Hoffman A (2001) Effects of recording media composition on the response of Nafion-coated carbon fiber microelectrodes measured using high-speed chronoamperometry. J Neurosci Methods 109:13–21
5. Crespi F, Sharp T, Maidment N et al (1983) Differential pulse voltammetry *in vivo* evidence that uric acid contributes to the indole oxidation peak. Neurosci Lett 43:203–207
6. Crespi F, Garratt JC, Sleight AJ et al (1990) In vivo evidence that 5-hydroxytryptamine (5-HT) neuronal firing and release are not necessarily correlated with 5-HT metabolism. Neuroscience 35:139–144
7. Cespuglio R, Sarda N, Gharib A et al (1986) Differential pulse voltammetry *in vivo* with working carbon fiber electrodes: 5-hydroxyindole compounds or uric acid detection? Exp Brain Res 64:589–595
8. Rivot J-P, Cespuglio R, Puig S et al (1995) *In vivo* electrochemical monitoring of serotonin in spinal dorsal horn with nafion-coated multi-carbon fiber electrodes. J Neurochem 65:1257–1263
9. Perez XA, Andrews AM (2005) Chronoamperometry to determine differential reductions in uptake in brain synaptosomes from serotonin transporter knockout mice. Anal Chem 77:818–826
10. Bunin MA, Wightman RM (1998) Quantitative evaluation of 5-hydroxytryptamine (serotonin) neuronal release and uptake: an investigation of extrasynaptic transmission. J Neurosci 18:4854–4860
11. Clements JD (1996) Transmitter timecourse in the synaptic cleft: its role in central synaptic function. Trends Neurosci 19:163–171

12. Cragg SJ, Rice ME (2004) DAncing past the DAT at a DA synapse. Trends Neurosci 27:270–277
13. Franklin KBJ, Paxinos G (1997) The mouse brain in stereotaxic coordinates. Academic, San Diego
14. Gulley JM, Doolen S, Zahniser NR (2002) Brief, repeated exposure to substrates down-regulates dopamine transporter function in Xenopus oocytes *in vitro* and rat dorsal striatum *in vivo*. J Neurochem 83:400–411
15. Gulley JM, Zahniser NR (2003) Rapid regulation of dopamine transporter function by substrates, blockers and presynaptic receptor ligands. Eur J Pharmacol 479:139–152
16. Blakely RD, Ramamoorthy S, Schroeter S et al (1998) Regulated phosphorylation and trafficking of antidepressant-sensitive serotonin transporter proteins. Biol Psychiatry 44: 169–178
17. Blakely RD, Bauman AL (2000) Biogenic amine transporters: regulation in flux. Curr Opin Neurobiol 10:328–336
18. Ramamoorthy S, Blakely RD (1999) Phosphorylation and sequestration of serotonin transporters differentially modulated by psychostimulants. Science 285:763–766
19. Daws LC, Toney GM, Davis DJ et al (1997) *In vivo* chronoamperometric measurements of extracellular serotonin clearance in the rat dentate gyrus. J Neurosci Methods 78:139–150
20. Daws LC, Toney GM, Gerhardt GA et al (1998) *In vivo* chronoamperometric measures of extracellular serotonin clearance in rat dorsal hippocampus: Contributions of serotonin and norepinephrine transporters. J Pharmacol Exp Ther 286:967–976
21. Daws LC, Callaghan PD, Morón J et al (2002) Cocaine increases dopamine uptake and cell surface expression of dopamine transporters. Biochem Biophy Res Commun 290: 1545–1550
22. Daws LC, Montañez S, Owens WA et al (2005) Transport mechanisms governing clearance *in vivo* revealed by high-speed chronoamperometry. J Neurosci Methods 143: 49–62
23. Daws LC, Montañez S, Munn L et al (2006) Ethanol inhibits clearance of brain serotonin by a serotonin transporter independent mechanism. J Neurosci 26:6431–6438
24. Daws LC, Toney GM (2007) Voltammetric methods to study kinetics and mechanisms for serotonin clearance *in vivo*. In: Michael AC, Simon SA, Nicolelis MAL (eds) Electrochemical methods in neuroscience, for Methods and new frontiers in neuroscience. CRC, Boca Raton
25. Baganz NL, Horton RE, Calderon AS et al (2008) Organic cation transporter 3: keeping the brake on extracellular serotonin in serotonin transporter deficient mice. Proc Natl Acad Sci U S A 105:18976–18981
26. Wiedholz L, Owens WA, Horton RE et al (2008) Mice lacking the AMPA GluR1 receptor exhibit hyperdopaminergia and "schizophrenia-related" behaviors. Mol Psychiatry 13:631–640
27. Baganz NL, Horton RE, Martin KP et al (2010) Repeated swim impairs serotonin clearance via a corticosterone-sensitive mechanism: organic cation transporter 3, the smoking gun. J Neurosci 30:15185–15195
28. Horton RE, Apple DM, Owens WA et al (2013) Decynium-22 enhances SSRI-induced antidepressant effects in mice: uncovering novel targets to treat depression. J Neurosci 33:10534–10543
29. Garcia-Olivares J, Torres-Salazar D, Owens WA et al (2013) Inhibition of dopamine transporter activity by a direct interaction with G protein βγ subunits. PLoS One 8(3):e59788
30. Montañez S, Munn JL, Owens WA et al (2014) 5-HT_{1B} receptor modulation of the serotonin transporter *in vivo*: studies using KO mice. Neurochem Int 73:127–131
31. Hoffman AF, Gerhardt GA (1999) Differences in pharmacological properties of dopamine release between substantia nigra and striatum: an *in vivo* electrochemical study. J Pharmacol Exp Ther 289:455–463
32. Callaghan PD, Irvine RJ, Daws LC (2005) Differences in the *in vivo* dynamics of neurotransmitter release and serotonin uptake after acute para-methoxyamphetamine and 3,4-methylenedioxymethamphetamine revealed by chronoamperometry. Neurochem Int 47:350–361
33. Fog JU, Khoshbouei H, Holy M et al (2006) Calmodulin kinase II interacts with the dopamine transporter C-terminus to regulate amphetamine-induced reverse transport. Neuron 51:417–429
34. Williams JM, Owens WA, Turner GH et al (2007) Hypoinsulinemia regulates amphetamine-induced reverse transport of dopamine. PLoS Biol 5:2369–2378
35. Robertson SD, Matthies HJG, Owens WA et al (2010) Insulin signaling regulation of norepinephrine transporter (NET) surface availability and function, reveals Akt as a novel and potent regulator of the transporter. J Neurosci 30:11305–11316
36. Speed N, Owens WA, Saada S et al (2011) Diet-induced changes in insulin signalling regulates

the trafficking and function of the dopamine transporter. PLoS Biol 6(9):e25169–e25169
37. Owens WA, Williams JM, Saunders C et al (2012) Rescue of dopamine transporter function in hypoinsulinemic rats by a D2 receptor-ERK dependent mechanism. J Neurosci 32:2637–3647
38. Rickhag M, Owens WA, Winkler M-T et al (2013) Membrane permeable C-terminal dopamine transporter peptides attenuate amphetamine-evoked dopamine release. J Biol Chem 288:27534–27544
39. Sabeti J, Adams CE, Burmeister J et al (2002) Kinetic analysis of striatal clearance of exogenous dopamine recorded by chronoamperometry in freely-moving rats. J Neurosci Methods 121:41–52
40. Jackson BP, Dietz SM, Wightman RM (1995) Fast-scan cyclic voltammetry of 5-hydroxtryptamine. Anal Chem 67:1115–1120
41. Jackson BP, Wightman RM (1995) Dynamics of 5-hydroxytryptamine released from dopamine neurons in the caudate putamen of the rat. Brain Res 674:163–166

Chapter 5

Experimental Data Guided Docking of Small Molecules into Homology Models of Neurotransmitter Transporters

Andreas Jurik, Amir Seddik, and Gerhard F. Ecker

Abstract

Docking of small molecules into proteins is a key process in structure-based drug design. It helps to derive binding hypotheses and is a standard tool for structure-based virtual screening. However, in case there are no high resolution structures of the protein of interest available, the results obtained require careful validation. In this chapter we present a workflow for experimental data guided docking of small molecules into protein homology models of neurotransmitter transporter.

Key words Neurotransmitter transporter, GABA transporter, Ligand docking, Scoring function, GAT1, SERT, DAT

1 Introduction

Molecular docking is a computational technique to obtain hypotheses for binding conformations of a ligand in a protein pocket. The method originated in the USA in the 1980s with a software package named DOCK, followed by other packages including also online services such as the ZDOCK server. At present times, there are numerous docking software packages available, which are typically run on a local personal computer or a small server. They vary in the sampling of the conformational space of the ligand and the protein, their algorithms for the placement of the ligand, and the type of scoring function for final ranking of the docking poses. On principle there are two main scenarios ligand docking is used for:

- Docking of one compound in order to derive a binding hypothesis for this compound. In this case multiple poses for the same compound are generated and compared in order to retrieve the energetically most favorable one.

Heinz Bönisch and Harald H. Sitte (eds.), *Neurotransmitter Transporters: Investigative Methods*, Neuromethods, vol. 118, DOI 10.1007/978-1-4939-3765-3_5,

- Docking of a large library of compounds and rank them according to their quality of fit, which is expected to correlate to their binding affinity. For this, only one pose per compound is considered, and different compounds are compared.

In both cases the use of the "right" scoring function is quite important, as different scoring functions quite often lead to different rankings of poses for the same compound. This is less problematic if a high resolution structure of the protein of interest co-crystallized with a high affinity ligand is available. In this case, usually a redocking experiment is performed in order to explore the best parameters for the docking runs. Briefly, the ligand is deleted from the binding site and docked again under various conditions using different scoring functions. Then the poses are compared with the original positioning of the ligand in the crystal structure on the basis of the rmsd of the heavy atoms. The setting for the closest match (usually an rmsd <2 Å is considered as appropriate) is then used for docking of new compounds. Unfortunately, in case of transmembrane transporters there are no high resolution structures of human analogs available. Hence, assessment of ligand binding modes by molecular docking studies into protein homology models of the respective transporter needs additional steps of validation in order to overcome the limitations of the different scoring functions. In this case most often consensus scoring, i.e., the use of multiple scoring functions followed by a voting scheme, is used. In light of our studies targeting the molecular basis of ligand-transporter interaction, we implemented a different protocol, which heavily relies on experimental data rather than energetic scoring functions for prioritizing docking poses. Briefly, a set of compounds from a homologous series is selected, which shows a pronounced structure-activity pattern. These compounds are docked into the protein and at least the top 100 poses are collected. Subsequently, the poses are clustered based on their common scaffold and the most populated clusters are analyzed. Those which are able to explain the structure-activity relationship pattern are finally prioritized and further validated. This approach has been successfully applied for docking of a set of benzodiazepines into the benzodiazepine binding site of the $GABA_A$ receptor [1] and for propafenone-type inhibitors of P-glycoprotein [2]. In this chapter we will outline the protocol for docking ligands into the binding site of SERT [3], DAT [4], NET [5], and GAT1 [6].

2 Materials and Methods

There are numerous commercial and open-source software packages for docking and scoring available. Genetic Optimization of Ligand Docking (GOLD) [7] is a commonly used commercial package wherein the user defines the pocket by a radius and a center by coordinates or by a reference ligand from a co-crystal structure. Hydrophobic and hydrogen bond fitting points on both the protein and the ligand are created. Optionally, side chain movements can be accounted for either by angles defined in a rotamer database [8] or a brute force exploration of all possible angles, typically in 10° bins. A genetic algorithm is employed whereby the ligand coordinates are converted into bit strings (also referred to as a chromosome). A scoring function is applied for each of the poses and two random poses are selected and weighted by their score. Of those two chromosomes, genetic operations are performed that typically include mutation, crossover, and migration "deepening" at a certain rate. The scoring is repeated until the desired number of poses has been reached. For the case study outlined below we used GOLD running on a dual octa core PC. The GOLD software has several scoring functions implemented, such as GoldScore, ChemScore, and ChemPLP. For common scaffold clustering XLSTAT [9] was used. Poses were visualized with MOE [10] or with Maestro [11].

3 Procedures

3.1 Ensemble Docking: Elucidating the Binding Event of GABA Reuptake Inhibitors

Investigating inhibitor binding in GAT1 is a procedure involving several distinct steps (Fig. 1). The binding event of lipophilic aromatic inhibitors is assumed to take place in an outward-open receptor conformation in order to accommodate the bulky ligands. As this significantly increases the number of potentially involved residues, protein flexibility can no longer efficiently be sampled by the use of rotamer libraries due to combinatorial explosion. Hence, a possible approach to explore the conformational flexibility of models for NSS members is to extract a set of diverse snapshots from a molecular dynamics simulation of a protein of interest and use it for docking.

3.1.1 Receptor Preparation

An early stable trajectory of an MD simulation should be used for the extraction of snapshots. This means making a compromise by avoiding early strained conformations and not going away too far from initial topologies. If a stable backbone RMSD is achieved after 10 ns of simulation, ten snapshots showing the highest RMSD within a radius of 7 Å around the ligand are extracted from the following 20 ns of the trajectory, followed by a soft energy minimization procedure covering a maximum of 5000 steps steepest descent and 50,000 steps of conjugate gradient minimization. Based on the occupancy in the binding site, presumably important water molecules can be additionally identified and kept for docking. This is achieved by analysis of the trajectory using either GROMACS [12] tools or within viewers like VMD [13].

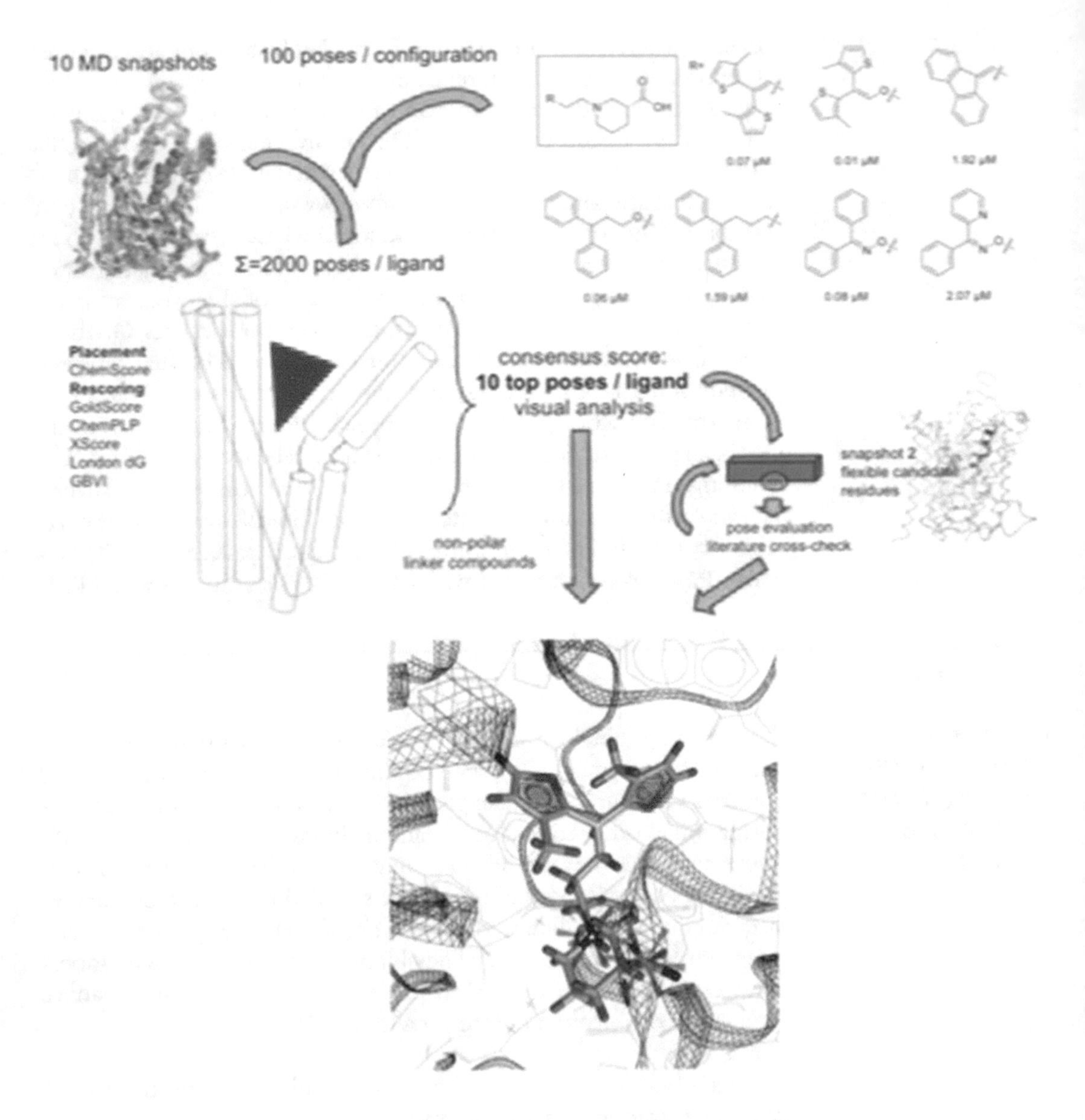

Fig. 1 Workflow for experimental data guided docking of tiagabine analogs into a series of homology models of GAT1 derived from molecular dynamics simulations

3.1.2 Ligand Preparation

Molecules used for docking need to be drawn (e.g. using the Molecular Operating Environment (MOE) software package) and processed for generating reasonable starting geometries, which can be done with CORINA [14]. Protonation states are sampled according to a physiological pH range of 7.2 ± 0.2 using LigPrep [15]. These states should be cross-checked with the major micro-species calculated by the ChemAxon web tool chemicalize.org.

3.1.3 Docking

GOLD provides an option for keeping present water molecules optional during docking and can be easily scripted, which comes in handy when dealing with multiple protein structures. However, careful inspection of input structures is essential (*see* **Note 2**). Providing ten MD snapshots at identical settings to the dataset to be docked, allowing for 100 poses per ligand and snapshot, is able to generate a sufficient amount of poses. Doing so, "early termination" and "diverse pose" options should be turned off in order to avoid bias to the pool of conformations.

3.1.4 Scoring

Due to the absence of crystallographically resolved, closely related protein-inhibitor complexes, redocking procedures are not an option for choosing a well-performing scoring function at this target. Hence, consensus scoring can provide improved performance over default or randomly chosen scoring algorithms. However, as sodium is present in the central binding cavity, ChemScore is a suitable primary scoring function, as it is also parameterized for metal interactions.

GOLD provides the possibility for rescoring poses with internally available scoring functions. For external rescoring with XScore, poses have to be exported to separate files, fitted with an appropriate identifier covering snapshot, ligand ID, and pose number. MOE is able to efficiently assign such identifiers, but also to rescore poses. Combining different types of scoring functions does not allow taking the numerical values into account. Rank-by-rank consensus scoring normalizes individual scoring values to indices, which are subsequently summed up for the determination of the common vote of scoring functions. This requires building up a database covering all docking poses with respective scoring values.

3.1.5 Pose Evaluation

Although the workflow explores a large area of conformational space on protein and ligand side, visual analysis of the ten best scored poses per ligand according to the ensemble of employed scoring functions quickly gives insights into the diversity of the yielded poses. This can be assisted but not replaced by automated analysis tools like the calculation of protein-ligand interaction fingerprints (PLIF). If multiple diverse binding modes are indicated, a larger population needs to be examined and clustered. Here, experiment-derived knowledge comes into play again to help eliminating unprobable binding poses or even pose clusters. For instance, if systematic sampling algorithms place the two bulky substituents of the tiagabine piperidine scaffold in energetically unfavorable axial orientations, according poses can be discarded, if no strong energetic benefit can be deduced from the bound complex. In case a single common binding mode for the whole used dataset can be identified, subsequent validation can be achieved by stressing against available structure-activity relationship (SAR) or prospective virtual screening campaigns.

4 Conclusions

Creating validated binding hypotheses of small molecules at neurotransmitter transporters by docking and scoring allows the design of new, more potent and selective ligands. However, due to the lack of high resolution structures the process needs careful experimental validation. Using common scaffold clustering combined with data from structure-activity relationship studies as constraints for prioritizing of docking poses is one possible approach. This has been successfully implemented for various transmembrane transporters, including GAT1, SERT, DAT, and the ABC-transporter P-glycoprotein. To account also for the conformational plasticity of these proteins, further developments will focus on incorporation of molecular dynamics simulations.

5 Notes

1. Use of "placeholders" for MD simulations

   In the presence of solutes, outward-open NSS models can show spontaneous closing of gating residues. In case that the dataset for subsequent docking studies comprises just one basic chemical scaffold, a "placeholder" ligand docked into the binding pocked can help to prevent such closure events. Furthermore, focused sampling of the binding site topology can be achieved.

2. PDB syntax issues

   Exporting and re-importing of PDB-formatted structures between different programs can have unexpected consequences. For instance, some programs differ in a single position to define the element of an atom. If another program now corrects for hydrogen atoms and misses the second position of an element, a sodium- (NA) and chloride (CL) containing binding site ends up filled with ammonia (N) and methane (C). While problems with standard proteins elements likely show up during the setup of a docking run, ions included in a binding event should always be checked separately for correct atom typing and charges. Corrections can be made relatively straightforward using a stream editor like sed.

Acknowledgments

We gratefully acknowledge financial support provided by the Austrian Science Fund, grant F3502.

References

1. Richter L, de Graaf C, Sieghart W et al (2012) Diazepam-bound GABAA receptor models identify new benzodiazepine binding-site ligands. Nat Chem Biol 8:455–464
2. Klepsch F, Chiba P, Ecker GF (2011) Exhaustive sampling of docking poses reveals binding hypotheses for propafenone type inhibitors of P-glycoprotein. PLoS Comput Biol 7:e1002036
3. Sarker S, Weissensteiner R, Steiner I et al (2010) The high-affinity binding site for tricyclic antidepressants resides in the outer vestibule of the serotonin transporter. Mol Pharmacol 78:1026–1035
4. Saha K, Partilla JS, Lehner KR et al (2015) "Second generation" mephedrone analogs, 4-MEC and 4-MePPP, differently affect monoamine transporter function. Neuropsychopharmacology 40:1321–1331
5. Neudorfer C, Seddik A, Shanab K et al (2015) Synthesis and *in silico* evaluation of novel reference compounds for PET based investigations of the norepinephrine transporter. Molecules 20:1712–1730
6. Jurik A, Zdrazil B, Holy M et al (2015) A binding mode hypothesis of tiagabine confirms liothyronine effect on γ-aminobutyric acid transporter 1 (GAT1). J Med Chem 58: 2149–2158
7. Jones G, Willett P, Glen RC et al (1997) Development and validation of a genetic algorithm for flexible docking. J Mol Biol 267:727–748
8. Lovell SC, Word JM, Richardson JS et al (2000) The penultimate rotamer library. Proteins 40:389–408
9. XLSTAT statistical add-in for Microsoft Excel, Addinsoft, 28 West 27th Street, Suite 503, New York, NY 10001, United States; www.xlstat.com
10. Molecular Operating Environment (MOE) version 2012.10 (2012) Chemical Computing Group Inc., 1010 Sherbrooke St. West, Suite #910, Montreal, QC, Canada, H3A 2R7
11. Schrödinger, 8910 University Center Lane, Suite 270, San Diego, CA 92122; www.schrodinger.com/maestro
12. Van Der Spoel D, Lindahl E, Hess B et al (2005) GROMACS: fast, flexible, and free. J Comput Chem 26:1701–1718
13. Humphrey W, Dalke A, Schulten K (1996) VMD: visual molecular dynamics. J Mol Graph 14:33–38
14. Sadowski J, Gasteiger J, Klebe G (1994) Comparison of automatic three-dimensional model builders using 639 X-ray structures. J Chem Inf Comput Sci 34:1000–1008
15. LigPrep, version 2.5 (2011) Schrödinger, LLC, New York, NY

Chapter 6

Immunohistochemical Methods for the Study of the Expression of Low-Affinity Monoamine Transporters in the Brain

Franck Louis, Thomas Couroussé, and Sophie Gautron

Abstract

Immunohistochemical and molecular methods, such as indirect immunofluorescence and in situ hybridization, have proven irreplaceable in providing an accurate vision of the precise localization of transporters as they occur within their native anatomical context in mammalian central nervous system, an essential prerequisite for understanding their function in vivo. Low-affinity monoamine transporters like organic cation transporters (OCTs) and plasma membrane monoamine transporter (PMAT) represent important yet poorly characterized components of aminergic neurotransmission in the brain. Contrarily to the classical high-affinity reuptake transporters, these low-affinity monoamine transporters are not restricted to the presynaptic terminals, but distributed diffusely in cells of numerous brain areas, in particular of aminergic projection regions. In this chapter, we will provide a detailed description of well-established procedures for diaminobenzidine (DAB) immunohistochemistry and immunofluorescence successfully applied to rodent brain. We also detail immunohistochemistry in situ hybridization co-labeling, a valuable approach which couples the sensitive detection of the cells expressing transporter mRNAs in brain sections with the immunohistochemical identification of these cells. We discuss conditions that can improve transporter detection and identify critical methodological steps in these protocols.

Key words Immunohistochemistry, Immunofluorescence, Low-affinity monoamine transporters, Central nervous system

1 Introduction

Developed in the early seventies, immunohistochemical and molecular methods such as indirect immunofluorescence and in situ hybridization have proven invaluable in providing the basic knowledge concerning the distribution and precise localization of neurotransmitter transporters in the central nervous system. Although conventional, these methods using tissue sections or whole mounts are the sole approaches allowing to get an accurate vision of the fine regional, cellular, and subcellular localization of transporters as they

Heinz Bönisch and Harald H. Sitte (eds.), *Neurotransmitter Transporters: Investigative Methods*, Neuromethods, vol. 118, DOI 10.1007/978-1-4939-3765-3_6, © Springer Science+Business Media New York 2016

occur within their native anatomical context in mammalian tissues, an essential prerequisite for understanding their function in vivo.

Low-affinity monoamine transporters like organic cation transporters (OCTs) and plasma membrane monoamine transporter (PMAT) represent novel and exciting targets for mood disorders, but have been altogether less studied by neuroscientists than the better-known high-affinity reuptake transporters. The high-affinity reuptake transporters exert an important control on aminergic neurotransmission, by ensuring the rapid reuptake of the released neurotransmitters (dopamine, serotonin, norepinephrine, and epinephrine) into the presynaptic terminals and their storage in synaptic vesicles. In the brain, these high-affinity transporters are found almost exclusively in the neurons that contain their cognate neurotransmitter and concentrated at high density in the plasma membrane of the synaptic and perisynaptic regions of nerve endings. In contrast, the three main OCTs (OCT1, OCT2, and OCT3) and PMAT, which belong to distinct families (Slc22 and ENT/Slc29, respectively), transport monoamines with much lower affinities, in the millimolar–submillimolar range, and exhibit a radically different distribution in the brain [1–8].

Anatomical studies during the last decades have demonstrated that these low-affinity monoamine transporters are not restricted to the presynaptic terminals as the high-affinity transporters, but distributed diffusely in cells of numerous brain areas, in particular of aminergic projection regions. To worsen the situation, these low-affinity monoamine transporters seem to be expressed at medium to low levels in the brain, PMAT and OCT3 messenger RNAs being more easily detected than that of OCT2, and OCT1 undetectable [9, 10]. In view of this diffuse localization, commonly used immunohistochemical methods were optimized and adapted to allow their efficient detection and characterization, and revealed a distinct expression pattern for each low-affinity transporter subtype. These methods present specific advantages and disadvantages, which are summarized in Table 1. Immunofluorescence visualizes the molecules of interest through detection of fluorochromes directly coupled to primary antibodies, or alternatively to secondary antibodies directed against the primary antibodies. The fluorochromes in the samples are excited with illumination at specific wavelengths, leading to emission at a longer wavelength range that is detected with a fluorescence microscope. Diaminobenzidine (DAB) immunohistochemistry uses avidin or streptavidin/biotin amplification of primary antigen–antibody complexes followed by chromogenic detection. Biotinylated horseradish peroxidase (HRP) oxidizes the chromogenic substrate DAB to yield a brown staining that can be analyzed with light microscopy. Conventional immunofluorescence studies using cryostat sections of paraformaldehyde-fixed brains allowed to detect the expression of organic cation transporter 3, a polyspecific low-affinity monoamine

Table 1
Advantages and disadvantages of different immunochemical methods for low-affinity monoamine transporter detection

Method	Advantages	Disadvantages	References
Immunohistochemistry	Allows regional overview at low magnification Good sensitivity	Lower intracellular resolution Potential leakage of signal	[2, 14, 16]
Immunofluorescence	Allows identification of cell type and organelles at high magnification Good intracellular resolution	Lower sensitivity Poor general overview at low magnification Photobleaching	[11–13, 15]
Immunohistochemistry/in situ hybridization co-labeling	Allows cell type identification	Lower intracellular resolution Long procedure	[10]
Immunofluorescence/in situ hybridization co-labeling	Allows cell type identification Rapid procedure	Poor general overview at low magnification Photobleaching	

transporter, within specific brain areas in rodents and to some extent refine its cellular and subcellular localization in neuron and astrocyte subpopulations by co-labeling with antibodies against specific markers of these cell types or compartments [11–13]. A full-blown determination of distribution of this transporter within specific neuroanatomical regions and systems was achieved by DAB immunohistochemistry using free-floating rat brain sections [14]. These studies showed that OCT3 was found principally in neuronal perikarya in multiple brain regions, including regions innervated by aminergic fibers, but also occasionally in fibers or in astrocytes in selected brain areas. Similar immunohistochemical and fluorescent approaches were also useful to define the expression profile of OCT2 in the brain, identifying as for OCT3 multiple sites of expression, with however a distinct profile from OCT3. DAB immunohistochemistry was used to reveal OCT2 expression in cell somas of human brain cortex and hippocampus [2]. Immunofluorescent labeling was used in later studies to establish OCT2 expression in neurons of aminergic nuclei such as dorsal raphe and locus coeruleus, and of several aminergic projection regions in mice brain [15]. Finally, these methods were used to identify brain regions and cell types expressing PMAT. DAB immunohistochemistry on mouse brain sections fixed with acetone was used to gain an overall picture of the brain distribution of this transporter [16], while radioactive PMAT in situ hybridization coupled to immunohistochemistry was used to refine the identification of the neuronal subtypes and systems expressing it [10]. In this chapter, we will recapitulate in detail the methods that our team has been using for over a decade for the anatomical characterization of OCT2, OCT3, and PMAT in rodent brain. Variables that can be controlled to improve sensitivity and specificity are summarized at several steps of the procedures.

2 Materials

2.1 Immuno histochemistry

2.1.1 Perfusion, Fixation, and Tissue Preparation

1. Animals: young adult animals are used (Sprague–Dawley rats between 3 and 4 months and C57BL6 mice between 2.5 and 4 months). *Animals develop with aging lipofuscin inclusions which can interfere with immunofluorescent detection of the transporters.*
2. Perfusion set-up: peristaltic perfusion pump (Gilson's MINIPULS 3) with 21G hypodermic needle attached for intracardial perfusion.
3. Microdissection instruments: fine forceps, large surgical and small sharp scissors, clamp.
4. Histology staining dishes.
5. Platform rocker.
6. Pentobarbital sodium.
7. Perfusion buffer: 0.9% NaCl (wt/vol), 0.1% sodium nitrate ($NaNO_2$) saline solution.
8. Fixation buffer: 4% paraformaldehyde (PAF) in 0.1 M Sorensen phosphate buffer.
9. Sorensen phosphate buffer (0.1 M) pH 7.2: Prepare stock solutions of 0.2 M Na_2HPO_4 and NaH_2PO_4. Mix 360 mL of Na_2HPO_4 and 140 mL of NaH_2PO_4, adjust buffer pH by adding either of the solutions and adjust volume to 1 L.

2.1.2 Free Floating Sections

1. Leica VT1000 S vibrating blade microtome (Leica Microsystems Inc.).
2. 1× Phosphate Buffered Saline (PBS): 0.01 M phosphate buffer, 0.154 M NaCl, pH 7.4.
3. 4% low-melting agarose in 1× PBS.
4. Methyl cyanoacrylate glue.
5. 12 or 6-well plastic plates.
6. 1× PBS with 0.02% (wt/vol) sodium azide (NaN_3).

2.1.3 Antigen Retrieval

1. Positively charged SUPERFROST PLUS microscope slides.
2. 0.1 M Sorensen phosphate buffer.
3. Retrieval buffer: 0.01 M triNa citrate pH 6.
4. 95 °C water bath or microwave oven.

2.1.4 DAB Immunochemistry

1. Positively charged SUPERFROST PLUS microscope slides and coverslips.
2. Fresh 30% hydrogen peroxide (H_2O_2).
3. 0.1–2% H_2O_2 (wt/vol) in distilled water.
4. Blocking buffer: 1× PBS, 4% Bovine Serum Albumin (BSA), 0.1–0.3% Triton X-100.
5. Primary antibodies against marker of interest. *Antibodies dilution depends on the source and range typically between 1/250–1/2000 for purified IgG and 1/1000–1/4000 for crude antisera.*
6. Biotinylated secondary antibodies against IgG of the species in which the primary antibodies were produced. Typical 1 μg/mL dilutions may be used but depend on the manufacturer.
7. Avidin or streptavidin and biotinylated horseradish peroxidase complex. Specific protocol and dilutions for complex formation depends on reagents and manufacturers.
8. DAB (3,3′-Diaminobenzidine). DAB can be brought as a ready-to-use solution or prepared from tablets at a final concentration of 0.7 mg/mL in 1× PBS containing 0.02% fresh H_2O_2.
9. 0.1 M Tris–HCl pH 7.5 buffer made from a 1 M Tris–HCl pH 7.5 stock solution.
10. 1 M Tris–HCl pH 7.5. Dissolve 121 g of Tris base in 600 mL of DEPC-treated H_2O. Adjust pH to 7.5 with 5 M HCl. Adjust final volume to 1 L with DEPC-treated H_2O. Autoclave for 20 min at 120 °C. Readjust pH after 1 month.
11. Fume hood for manipulation of solvent.
12. Solutions of increasing EtOH concentration (70%, 95%, 2 baths of 100%, vol/vol) in distilled water.
13. Xylene.
14. Resin-based mounting medium (Eukitt).

2.1.5 Immuno-fluorescence

1. Positively charged SUPERFROST PLUS microscope slides and coverslips.
2. Blocking buffer: 1× PBS, 4% Bovine Serum Albumin (BSA), 0.1–0.3% Triton X-100.
3. 1× PBS.
4. Primary antibodies against marker of interest. *Antibodies dilution depends on the source and range typically between 1/250–1/2000 for purified IgG and 1/1000–1/4000 for crude antisera.*
5. Fluorescent (Alexa 488 or 555) secondary antibodies directed against IgG of the species in which the primary antibodies were produced. Typical 1 μg/mL dilutions may be used but depend on the manufacturer.

6. Fluoromount-G mounting medium (±DAPI stain 0.5 μg/mL) containing 0.1 % NaN_3.

2.2 In Situ Hybridization

Specific precautions must be taken to limit degradation of mRNAs by endogenous and exogenous RNAses. Disposable gloves must be worn during manipulation, RNAse-free glass or sterile disposable plastic ware, and ultrapure (Type 1) or diethylpyrocarbonate (DEPC)-treated water must be used for buffers. To degrade RNAses, glassware can be baked at 300 °C for 4 h or treated with 0.1 % (vol/vol) DEPC in water for 1 h followed by autoclaving for 20 min at 120 °C to destroy DEPC. DEPC is toxic and must be pipetted in a fume hood.

2.2.1 Tissue Preparation

1. Animals.
2. Dry ice.
3. Isopentane (2-methylbutane)
   Isopentane is highly flammable. Store in a cool and well-ventilated area. Do not manipulate at proximity of a flame.
4. 0.1 % $NaNO_2$ in physiological serum (0.9 % NaCl).
5. OCT embedding medium.
6. Plastic cryomolds.
7. 1× PBS.
8. Positively charged SUPERFROST PLUS microscope slides and coverslips.

2.2.2 Fixation of Sections

1. Fixative: 3.7 % formaldehyde (vol/vol) in 1× PBS.
2. DEPC-treated water: Add DEPC to H_2O to a concentration of 0.1 % (vol/vol), mix well by shaking and let stand the mixture in a fume hood overnight. Autoclave for 20 min at 120 °C to inactivate DEPC.
3. Solutions of increasing EtOH concentration (50 %, 70 %, two baths of 100 %, vol/vol) in distilled water.

2.2.3 Oligonucleotidic Probes Preparation

Labeling by Tailing

1. RNAse-free 1 mL Eppendorf tubes.
2. Terminal deoxynucleotidyl transferase buffer.
3. Oligonucleotides: Oligomeric DNA probes (45 nucleotides long) are selected based on their sequence specificity. With the help of computer programs, select oligonucleotides in specific regions of the mRNA of interest to avoid hybridization with unwanted targets, with a GC content between 48 and 62 % and avoiding hairpins and repeats. Use purified oligonucleotides. *To enhance detection of low-copy mRNAs, we used a mixture of three non-overlapping oligonucleotides.*
4. ^{35}S-dATP (1250 Ci (46.2 TBq)/mmol, 12.5 mCi/mL).

5. Terminal deoxynucleotidyl transferase. *The enzyme is sensitive to air thus must be added at the end when setting up the reaction. Excess DTT is harmful since it can interact with cacodylate in the tailing buffer.*
6. 0.5 M EDTA pH 7.2: Dissolve 73.1 g EDTA in 250 mL H_2O. Add 5 M NaOH progressively until the EDTA dissolves and reaches pH 8. Bring up to 500 mL with DEPC-treated H_2O. Sterilize by autoclaving for 20 min.

Probe Purification

1. Sephadex G-25 microspin columns.
2. 2 M DTT: Dissolve 1.5 g of DTT in DEPC-treated H_2O. Adjust the volume to 10 mL and store as 1 mL aliquots in the dark (wrapped in aluminum foil) at −20 °C. DTT or DTT-containing solutions must not be autoclaved.

2.2.4 Hybridization

1. A shaking water bath.
2. 20× SSC buffer pH 7 (3 M NaCl, 0.3 M triNa citrate): Add 175.3 g NaCl and 77.4 g triNa citrate to 800 mL of DEPC-treated H_2O. Adjust the pH to 7.0 with a few drops 10 N HCl. Complete volume to 1 L. Autoclave and store at room temperature.
3. Denhardt's solution (100×): Dissolve 1 g Polyvinylpyrrolidone (PVP), 1 g Ficoll 400, 1 g Glycine in H_2O, complete to 50 mL DEPC-treated H_2O.
4. Salmon DNA (10 mg/mL): Dissolve 1 g of salmon or herring sperm DNA in 100 mL DEPC-treated H_2O, autoclave for 20 min and store at −20 °C.
5. Deionized formamide.
6. Polyadenylic acic (poly[A]) 10 mg/mL in DEPC-treated H_2O.
7. Hybridization buffer: 2× SSC, 1× Denhardts solution, 300 μg/mL salmon sperm DNA, 40% deionized formamide, 0.1 M DTT, 0.5 mg/mL PolyA and 4×10^6 dpm/mL of each labeled sense or antisense oligonucleotides. *Care should be taken that all components are well dissolved, to avoid high background and nonspecific signals.*
8. Plastic slide cases (mailers).

2.2.5 Immunohistochemistry

1. Fresh 30% hydrogen peroxide (H_2O_2).
2. Blocking buffer: 1× PBS, 6% normal horse serum, 0.1–0.3% Triton X-100.
3. Primary antibodies against marker of interest. *Antibodies dilution depends on the source and range typically between 1/200–1/2000 for purified IgG and 1/1000–1/4000 for crude antisera.*
4. Biotinylated secondary antibodies against IgG of the species in which the primary antibodies were produced. Typical 1 μg/mL dilutions may be used but depend on the manufacturer.

5. DAB at a final concentration of 0.7 mg/mL in 1× PBS containing 0.02 % fresh H_2O_2.
6. 0.1 M Tris–HCl pH 7.5 buffer made from a 1 M Tris–HCl pH 7.5 stock solution.
7. 1 M Tris–HCl pH 7.5: Dissolve 121 g of Tris base in 600 mL of DEPC-treated H_2O. Adjust pH to 7.5 with 5 M HCl. Adjust final volume to 1 L with DEPC-treated H_2O. Autoclave for 20 min at 120 °C. Readjust pH after 1 month.

2.2.6 Emulsion and Revelation

1. A light-proof dark room with a water bath.
2. A glass staining jar slightly above the size of slides.
3. Emulsion (Ilford K5, Kodak, Rochester, NY).
4. Desiccant/silica gel.
5. Developer (D19 Kodak) diluted (vol/vol, 1:1) with distilled H_2O.
6. Fixative (Kodak).

3 Methods

3.1 Immuno-histochemistry

3.1.1 Tissue Preparation

1. Deeply anesthetize mice or rat with pentobarbital sodium (30–70 mg/kg i.p.). Progression of anesthesia must be monitored by evaluation of corneal and toe pinch reflexes and jaw tone.
2. With the help of the peristaltic pump, transcardially perfuse the animals with 20 mL (60 mL for rat) physiological serum containing 0.1 % $NaNO_2$ to favor vasodilatation, followed by 140 mL (400 mL for rat) freshly made ice-cold fixation buffer containing 4 % (wt/vol) PAF in 0.1 M Sorensen phosphate buffer. *Initial perfusion with physiological serum allows the elimination of haematids. The perfusion flow must be high enough to drive efficient circulation of the fixative but not too rapid to avoid clogging and insufficient fixation. A volume flow of 10–15 mL per min is a typical range. Post-perfusion stiffness of the animals and discoloring of the organs are a good indication of the efficacy of fixation, which can have important bearings on epitope recognition in the subsequent steps.*
3. Immediately remove the brains and post-fix by overnight immersion in fixation buffer (50 mL) at +4 °C.
   The volumes given in these initial steps are for adult animals. PAF concentration and volumes must be adapted for younger animals. Immunochemistry may also work on frozen unfixed tissue, but introduction of fixation steps allows the preservation of the anatomical structure of the tissue. Several fixatives may be tested (ethanol, acetone, methanol, acetic acid). PAF, which is compatible with most epitopes and antibodies and provides excellent conservation of the fine tissue structure, is commonly used.

3.1.2 Free-Floating Sections

1. Imbed the brains in histology molds in 4% low-melting agarose in 1× PBS at 37 °C. After solidification of the agarose at room temperature, orient and fix the brain directly on the vibratome support using a small amount of methyl cyanoacrylate glue.
2. Cut 20–40 μm brain sections with the vibrating microtome. Sections can be stored in 12- or 6-well plates and kept up to 2 weeks at +4 °C in 1× PBS containing NaN_3 (0.02%, wt/vol). Tissue sections are processed either for DAB or fluorescent immunochemistry. *The use of free-floating sections improves the sensitivity of detection of low-affinity transporters since immunochemical reagents can penetrate more easily the tissues. While thick sections are easier to manipulate, they show a higher level of nonspecific background. NaN3 must be added for storage over 2 days, but may occasionally interfere with epitope recognition. Moreover, in our hands, efficacy of detection of some transporters was found to decrease significantly after storage over longer periods of time of vibratome sections stored in a cryoprotectant solution at −20 °C as well as of cryostat sections kept frozen at −80 °C.*

3.1.3 Antigen Retrieval (Optional)

Antigen retrieval may improve significantly low-affinity monoamine transporter detection. The sections are mounted on glass slides to avoid collapse, which precludes use of free-floating sections in subsequent steps. *In this case, all steps are identical except that sections mounted on slides are incubated in dedicated boxes instead of free-floating.*

1. Mount the sections on Superfrost slides in 0.1 M Sorensen phosphate buffer, drain carefully excess buffer and dry on a hot plate (45 °C) until they adhere (approximately 1 min). Do not overdry.
2. Incubate the slides 5 min in retrieval buffer (0.01 M triNa citrate, pH 6), discard buffer, then incubate 20 min in fresh retrieval buffer at 95 °C in a water bath.
3. Alternately, microwave retrieval may be used on mounted sections. The slides must be immerged in retrieval buffer in a staining dish and microwave twice for 2.5 min (oven at ~700 W). Allow cooling for 20 min in microwave without opening.
4. Rinse twice in 1× PBS at room temperature.

3.1.4 Processing for Immunodetection

DAB Immunochemistry

1. To block endogenous peroxidase activity, incubate the sections 10 min in increasing concentrations of fresh H_2O_2 ranging from 0.1 to 2% until bubbles production stops. *H2O2 solution exposed to air deteriorates quickly so freshly opened bottles or aliquots kept frozen must be used. Plastic sieves of the size of the wells are used to drain sections from this step on.*

2. Wash sections in 1× PBS for 5 min under moderate agitation on a platform rocker.
3. Incubate sections for 30 min in blocking buffer (1× PBS, 0.3% triton, 4% BSA) under moderate agitation at room temperature agitation on a platform rocker. *Triton X-100 concentration must be tested to optimize detection. We found that high concentrations (0.3%) increased specific signal. Plastic sieves adjusted to the wells are used to drain sections from this step on. BSA can be replaced by normal antiserum from the species in which secondary antibodies were produced.*
4. Incubate sections overnight at +4 °C in blocking buffer containing appropriate dilution of primary antibodies or antiserum against low-affinity transporter. *Antibodies dilution depends on the source and range typically between 1/250–1/5000 for purified IgG and 1/2000–1/20000 for crude antisera.*
5. Wash sections three times for 10 min in an excess volume of 1× PBS at room temperature under mild agitation on platform rocker.
6. Incubate sections for 2 h with biotinylated secondary antibodies against IgG of the species in which the primary antibodies were produced. Typical 1 μg/mL dilutions may be used but depend on the manufacturer.
7. Wash sections three times for 10 min in an excess volume of 1× PBS at room temperature under mild agitation on platform rocker.
8. Incubate sections for 1 h with pre-formed (30 min) avidin or streptavidin and biotinylated horseradish peroxidase complex. *The avidin–biotin complex (ABC) method takes advantage of the properties of avidin from egg white and streptavidin from Streptomyces avidinii which each have four high-affinity binding sites for biotin. The advantages of commercial forms of streptavidin over avidin are a better affinity for biotin when conjugated, a near neutral isoelectric point and the absence of carbohydrate groups, resulting in minimal background staining. In this method, sections are incubated sequentially with primary antibodies, secondary antibodies conjugated to biotin, and finally large pre-formed avidin–biotin–horseradish peroxidase (HRP) complexes. This results in increased signal intensity and sensitivity, revealed by HRP oxidization of the chromogenic substrate DAB. Specific protocol and dilutions for complex formation depend on reagents and manufacturer.*
9. Wash sections three times for 10 min in an excess volume of 1× PBS at room temperature under moderate agitation on a platform rocker.
10. Incubate sections in peroxidase substrate DAB as specified by the manufacturer (usually 0.7 mg/mL in 1× PBS containing 0.02% fresh H_2O_2) while monitoring the progression of staining

up to 10 min. *DAB is classified as toxic thus it is recommended to use tablets or ready-to-use solutions instead of free powder and to refer to the specific hazard statements and safety instructions before using.*

11. Stop the enzymatic reaction by a 2 min incubation in Tris–HCl 0.1 M pH 7.5 buffer.
12. Rinse quickly (2 s) sections in tap water, mount on slides, and air dry. *Histological counterstaining may be done at this step if it does not interfere with DAB coloration.*
13. Dehydrate the sections with increasing EtOH concentrations (50%, 95%, and two baths of 100%) and air dry. Quickly dip the slides in xylene, mount under coverslips with a drop of mounting medium, and view under a light microscope (Fig. 1). Images can be captured using a digital camera (Carl Zeiss AxioCam) with appropriate software (Axiovision) and converted into TIFF format. Whole slide imaging, i.e., scanning of the glass slides and conversion into digital slides, significantly facilitates the analysis of the regional distribution.

Immunofluorescence

1. Incubate sections for 30 min in blocking buffer (1× PBS, 0.3% Triton X-100, 4% BSA) under moderate agitation on a platform rocker at room temperature. *Triton X-100 concentration must be tested to optimize transporter detection. We found that high concentrations (0.3%) increased specific signal. Plastic sieves adjusted to the wells are used to drain sections from this step on.*
2. Incubate sections overnight at +4 °C in blocking buffer containing appropriate dilution of primary antibodies or antiserum against low-affinity monoamine transporter. *Antibodies dilution depends on the source and range typically between 1/250–1/2000 for purified IgG and 1/1000–1/4000 for crude antisera.*

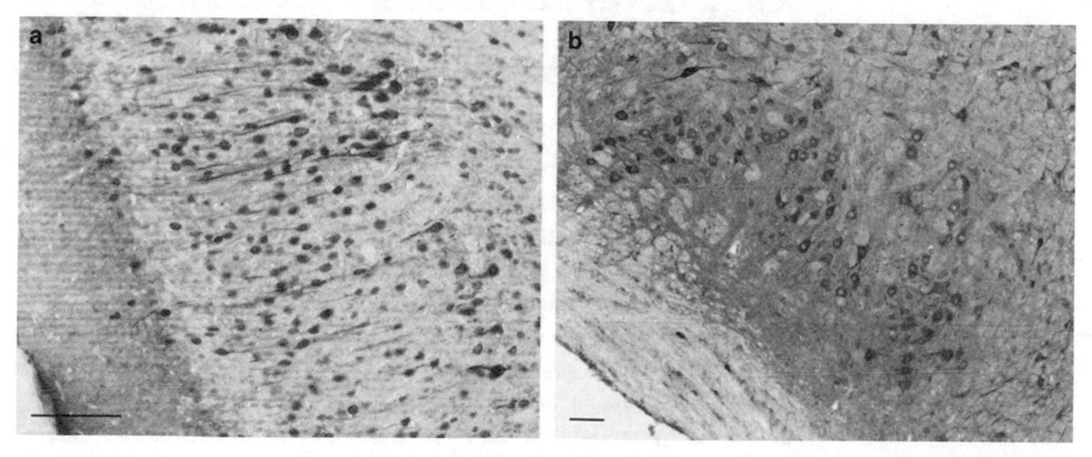

Fig. 1 DAB immunochemistry. Low-magnification light-field photomicrographs of coronal rat brain sections showing DAB immunostaining of OCT3 (×20) in neurons of retrosplenial granular cortex (**a**) and of PMAT (×10) in neurons of the rat facial nucleus (**b**). Scale bars represent 100 μm

3. Wash sections three times for 10 min in an excess volume of 1× PBS at room temperature under moderate agitation on platform rocker.
4. Incubate sections for 1 h with fluorescent secondary antibodies (for instance labeled with Alexa 488 or 555) directed against IgG of the species in which the primary antibodies were produced. Typical 1 μg/mL dilutions may be used but depend on the manufacturer. *Fluorescent chromophores are photosensitive thus photobleaching must be avoided by limiting time and intensity of exposure throughout the procedure and improved by choosing fluorochromes less prone to photobleaching.*
5. Wash sections three times for 10 min in an excess volume of 1× PBS at room temperature under moderate agitation on platform rocker.
6. Mount the sections on slides (do not dry) under coverslips with a drop of fluorescence mounting medium. *DAPI stain (at 0.5 μg/mL final concentration) may be added to the mounting medium to visualize cell nuclei on sections.* Images can be captured using a digital camera (Carl Zeiss AxioCam) with appropriate software (Axiovision) and converted into TIFF format. For multi-labeling, images are pseudo-colored and merged to visualize overlap (Fig. 2). As for DAB immunohistochemistry, whole slide imaging is highly recommended for the study of regional or cellular distribution with multi-labeling.

Colocalization experiments may be performed with primary antibodies against a marker of interest made in a distinct species, incubated simultaneously or sequentially, and revealed with specific secondary antibodies chosen such as to avoid species cross-reactivity. These secondary antibodies must be labeled with distinct fluorochromes, excited and emitting at distinguishable wavelengths (such as Alexa 488 and 555). Particular care must be taken in choosing narrow band-pass filters to avoid cross-excitation of fluorochromes and leakage by detection of emission of a given fluorochrome through the wrong filter.

In specific cases, problematic detection of scarcely expressed or poorly accessible molecules may benefit from an additional amplification step, such as amplification with streptavidin/biotin, or the tyramide signal amplification (TSA) method. In the first method, an additional amplification step is introduced by incubation with secondary antibodies conjugated with biotin then with fluorescent streptavidin. Secondary antibodies conjugated with several small-sized molecules of biotin each can bind streptavidin, greatly increasing the sensitivity of the assay. The second method, TSA, is an enzyme-mediated detection method that uses horseradish peroxidase (HRP) to generate high-density labeling at the vicinity of a target protein. The HRP molecule is conjugated to secondary antibodies and

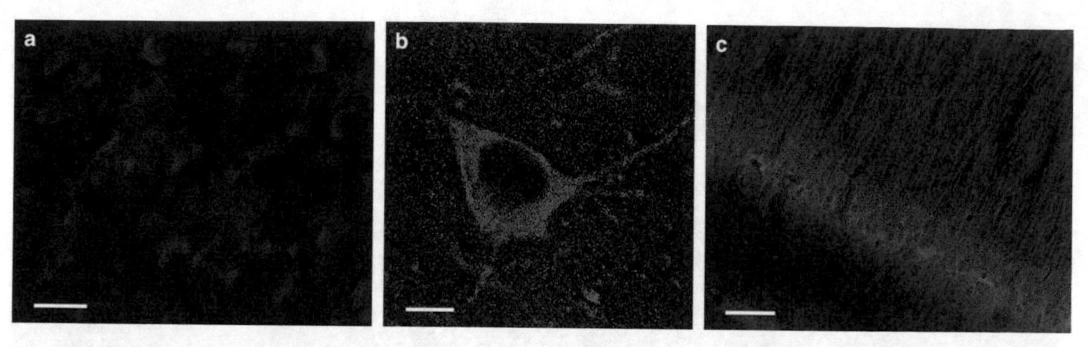

Fig. 2 Immunofluorescence labeling. Fluorescent photomicrographs (63×) of coronal mouse brain sections. Conventional high-magnification microscopy showing labeling of OCT3 in frontal cortex (**a**), confocal microscopy showing labeling of OCT2 in frontal cortex (**b**) and conventional high-magnification microscopy showing co-labeling of OCT2 (*green*) and neuronal marker NeuN (*red*) in neurons of hippocampus CA1 (**c**). Scale bars represent 10 μm (**a** and **b**) and 50 μm (**c**)

converts the fluorescent TSA reagent (i.e. Cyanine 3) to highly reactive free radicals that form covalent bonds with tyrosines near the enzyme. This method improves the detection sensitivity by 100–1000 over standard methods, but is associated with a loss in resolution.

Visualization and image acquisition can be carried out on a wide-range microscope to gain a general overview, with high magnification to analyze the signal at the cellular level. For subcellular localization, confocal microscopy which enables the reconstruction of three-dimensional images from a stack of focal planes may also be used to increase resolution and contrast (Fig. 3). Particular care must be taken to ensure that apparently overlapping elements are located in the same plane of the Z-axis (using for instance Volocity 3D Image Analysis Software from Perkin Elmer).

For both DAB immunohistochemistry and immunofluorescence, verification of the specificity of the primary antibodies can be done by pre-incubating these antibodies with increasing concentrations of the epitope against which they were raised (peptide or protein). These absorption tests do not exclude however the possibility of a cross-reaction with a common epitope in an unrelated molecule. Definitive proof of specificity can be obtained by testing the antibodies on tissues in which the molecule of interest is lacking, such as in knockout mice, or on cells overexpressing the molecule of interest. The specificity of secondary antibodies on the other hand can be tested on sections for which incubation with primary antibodies has been omitted or incubated with IgGs produced in non-target species.

3.2 Immunohistochemistry In Situ Hybridization Co-labeling

Due to the widespread expression profile and diffuse localization of low-affinity monoamine transporters in the brain, anatomical methods of study have been optimized to allow their efficient detection and characterization. Immunohistochemistry in situ hybridization co-labeling offers the advantage of combining

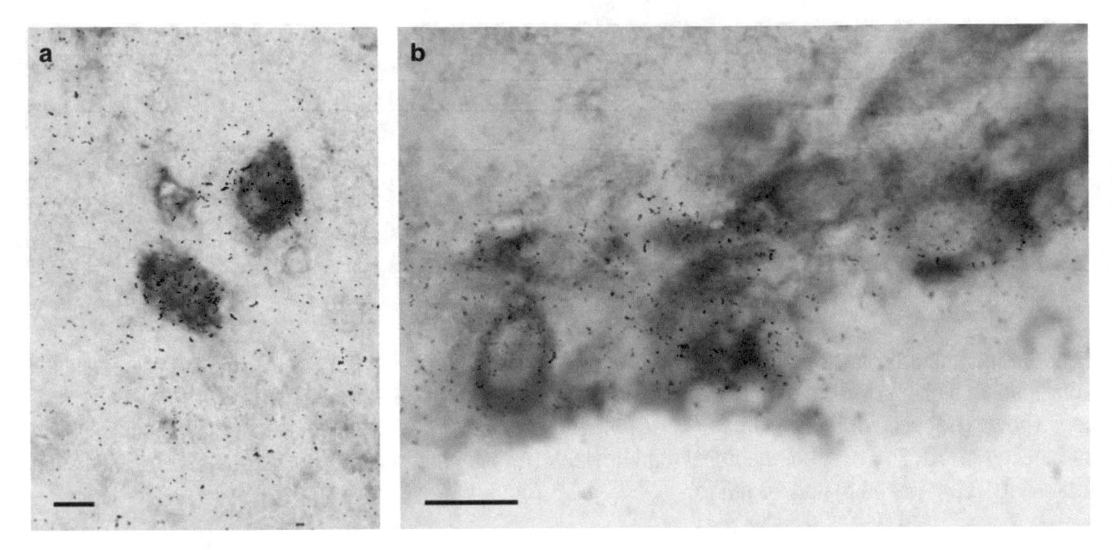

Fig. 3 PMAT in situ hybridization/immunochemistry co-labeling. High-magnification (63×) light-field photomicrographs show colocalization of PMAT mRNA in situ hybridization (silver grains) with DAB immunohistochemistry for tyrosine hydroxylase (TH) in the substantia nigra, pars reticulata (**a**) and for histidine decarboxylase (HDC) in the tuberomammillary nucleus (**b**). Scale bars represent 10 μm

sensitive radioactive detection of the mRNAs encoding the transporters at the regional and cellular levels with the immunochemical identification of the cell subtypes expressing these mRNAs. The use of multiple long oligonucleotides radiolabeled by tailing provides increased sensitivity of the in situ hybridization step.

3.2.1 Tissue Preparation

1. Sacrifice the animals by decapitation.
2. Quickly dissect the brain on ice, freeze them by immersion in 2-methylbutane cooled at −35 °C with dry ice and store at −80 °C. Cut 16 μm coronal or sagittal sections at −20 °C on a cryostat and thaw-mount onto poly-l lysine-coated glass slides. *The sections can then be directly fixed or stored frozen at −80 °C and fixed on the day of the experiment.*

3.2.2 In Situ Hybridization

Fixation of Sections

1. Fix sections in 3.7% formaldehyde in 1× PBS for 1 h at room temperature.
2. Wash 5 min in 1× PBS.
3. Rinse 10 s in DEPC-treated H_2O.
4. Dehydrate in baths (30 s) of increasing EtOH concentrations (50%, 70%, two baths at 100%).
5. Air-dry 30 min and either proceed with hybridization or store the slides dehydrated at −80 °C for subsequent utilization.

Oligonucleotidic Probes Preparation

Oligonucleotides (45-mer) are designed to detect specifically mRNAs encoding low-affinity transporters while minimizing hybridization with nonspecific targets. A mixture of three antisense oligonucleotides can be used to enhance detection and a mixture of the corresponding sense oligonucleotides are used as negative control. *The oligonucleotides are 3′ end-labeled with [α-35S]dATP at a final specific activity between 5.109 and 2.1010 cpm/μg in order to detect RNAs with diffuse and low expression levels such as those encoding the low-affinity transporters.*

Probe Labeling by Tailing

1. On ice, prepare a 100 μL of a mixture made by successive addition of DEPC-treated H_2O, 2 pmol of each oligonucleotide, 1× terminal deoxynucleotidyl transferase buffer, 2 μL of ^{35}S-dATP (1250 Ci or 46.2TBq/mmol, 12.5 mCi/mL) and 30 units of terminal deoxynucleotidyl transferase following the manufacturer's protocol.
2. Incubate 45 min at 37 °C.
3. Stop the reaction with 5 μL of 0.5 M EDTA pH 7.5.

Probe Purification

1. Separate the labeled oligonucleotides from unreacted with a Sephadex G-25 microspin column as specified by the manufacturer.
2. Measure the recovered elution volume after elution and add 1/10 vol of freshly prepared 2 M DTT in DEPC-treated H_2O.
3. Count by scintillation 1 μL of elution in order to evaluate the specific activity of the radiolabeled oligonucleotides.

Probe Hybridization

1. On the day of the experiment, immerge the sections in 70% EtOH immediately after removing from the deep-freeze and let dry for 30 min.
2. Prepare hybridization in buffer containing 2× SSC, 1× Denhardts solution, 300 μg/mL salmon DNA, 40% deionized formamide, 0.1 M DTT, 0.5 mg/mL PolyA and a mixture of labeled sense or antisense oligonucleotides (4×10^6 dpm/mL), vortex and briefly centrifuge the solution and maintain at 50 °C until incubated with section.
3. Drop 100–200 mL of probe mix per section and overlay with hydrophobic plastic coverslips of minimal size. Install flat in plastic slide mailers. Incubate for overnight at 42–50 °C in a humidified chamber. *Hybridization temperature must be determined depending on the calculated melting temperature (Tm) of oligonucleotides which depends on their length and GC content.*

Wash of Sections

1. Immerge the slides in 1× SSC and carefully lift off the coverslips using fine-tipped forceps.
2. Wash the slides successively in two baths of 1× SSC, 10 mM DTT, two baths of 0.5× SSC, 10 mM DTT in a shaking water

bath at 53 °C. *Particular care must be given to maintaining the sections damp.*

3.2.3 Immunohistochemistry

1. After the last wash step, rinse sections 5 min in 1× PBS and fix in 4% PAF in 1× PBS for 30 min. *This specific fixation step is crucial to protect the mRNAs from degradation.*
2. Wash sections three times for 10 min in an excess volume of 1× PBS at room temperature under mild agitation.
3. Incubate 1 h in blocking buffer (1× PBS, 6% heat-inactivated horse serum, 0.1–0.3% Triton X-100). *Horse serum contains usually less RNAse than other species, thus limiting mRNA degradation during this step. Triton concentrations must be tested and optimized, for instance choline acetyltransferase (ChAT) immunohistochemistry does not tolerate Triton X-100.*
4. Incubate sections 2 h at room temperature in blocking buffer containing appropriate dilutions of primary antibodies directed against marker of interest. *Antibodies dilution depends on the source and range typically between 1/250–1/1000 for purified IgG and 1/1000–1/4000 for crude antisera.*
5. Wash sections three times for 10 min in an excess volume of 1× PBS at room temperature under mild agitation.
6. Incubate sections for 1 h with biotinylated secondary antibodies against IgG of the species in which the primary antibodies were produced. Typical 1 μg/mL dilutions may be used but depend on the manufacturer.
7. Wash sections three times for 10 min in an excess volume of 1× PBS at room temperature under mild agitation.
8. Incubate sections for 1 h with pre-formed (30 min) avidin or streptavidin and biotinylated horseradish peroxidase complex. *Specific protocol and dilutions for complex formation depends on reagents and manufacturer.*
9. Wash sections three times for 10 min in an excess volume of 1× PBS at room temperature under mild agitation.
10. Incubate sections 10 min in DAB solution as specified by the manufacturer (usually 0.7 mg/mL in 1× PBS containing 0.02% fresh H_2O_2).
11. Stop the enzymatic reaction and rinse with 3× 10 min incubations in 0.1 M Tris–HCl pH 7.5 buffer.
12. Rinse quickly (2 s) sections in tap water, mount on slides, and air dry.
13. Dehydrate the sections with EtOH (70%, and 100%, 30 s each) and air dry.

3.2.4 Emulsion and Revelation

Emulsion

Work in a lightproof dark room at +15 °C, at a distance from a red light. The emulsion must be protected from light, heat, and radioactivity.

1. Prepare a water bath at 42 °C containing a water beaker.
2. Dilute emulsion (Ilford K5, Kodak, Rochester, NY) with 2× SSC (v/v, 1:1), or as specified by the manufacturer. Homogenize carefully without generating bubbles.
3. Melt the required volume of photographic emulsion in water bath at 42 °C. *20 mL allows emulsion of approximately 65 slides. Limit incubation time at 42 °C, which veils the emulsion.*
4. Pour carefully the photographic emulsion in a glass staining jar set up in the beaker in the water bath at 42 °C, without generating bubbles. *To optimize emulsion volume, use a jar slightly above slide size.*
5. Dip each slide 5 s in photographic emulsion up to the frosted end and draw out leaving in the initial upright position. While maintaining the slide position, drain excess emulsion, wipe the back of the slide with a tissue and install the slides on slide stand. Let dry in a light-proof box for 2–10 h in presence of desiccant (silica gel).
6. The following day, put the slides in a slide box with desiccant, protect from light by wrapping with two aluminum foils. Expose for 4–6 weeks.

Revelation

In dark room at +15 °C, at a distance from a red light, incubate the slides successively:

- 3 min in developer (D19 Kodak) diluted (v/v, 1:1) with distilled H_2O.
- 20 s in distilled H_2O.
- 5 min in fixative (Kodak).
- Three times for 5 min in distilled H_2O.

Optionally, the sections can at this step be counterstained with hematoxylin–eosin. Following development, sections are dehydrated through graded ethanol (see Sect. 3.2.3.13), mounted under coverslips and viewed under a light microscope (Fig. 3).

Acknowledgments

We gratefully acknowledge the contribution of several former team members for the development and application of the methods described, in particular Vincent Vialou, Alexandre Bacq, Laure Balasse, and Quentin Perrenoud. Financial support was provided by the Institut National de la Santé et la Recherche Médicale (Inserm) and Fondation de France and fellowships from the French Ministry for Research, the Fondation pour la Recherche Médicale, and the Société Française de Pharmacologie et Thérapeutique.

References

1. Breidert T, Spitzenberger F, Grundemann D, Schomig E (1998) Catecholamine transport by the organic cation transporter type 1 (OCT1). Br J Pharmacol 125:218–224
2. Busch AE, Karbach U, Miska D, Gorboulev V, Akhoundova A, Volk C et al (1998) Human neurons express the polyspecific cation transporter hOCT2, which translocates monoamine neurotransmitters, amantadine, and memantine. Mol Pharmacol 54:342–352
3. Grundemann D, Schechinger B, Rappold GA, Schomig E (1998) Molecular, identification of the corticosterone-sensitive extraneuronal catecholamine transporter. Nat Neurosci 1:349–351
4. Grundemann D, Koster S, Kiefer N, Breidert T, Engelhardt M, Spitzenberger F et al (1998) Transport of monoamine transmitters by the organic cation transporter type 2, OCT2. J Biol Chem 273:30915–30920
5. Grundemann D, Liebich G, Kiefer N, Koster S, Schomig E (1999) Selective substrates for non-neuronal monoamine transporters. Mol Pharmacol 56:1–10
6. Wu X, Kekuda R, Huang W, Fei YJ, Leibach FH, Chen J et al (1998) Identity of the organic cation transporter OCT3 as the extraneuronal monoamine transporter (uptake2) and evidence for the expression of the transporter in the brain. J Biol Chem 273:32776–32786
7. Engel K, Zhou M, Wang J (2004) Identification and characterization of a novel monoamine transporter in the human brain. J Biol Chem 279:50042–50049
8. Engel K, Wang J (2005) Interaction of organic cations with a newly identified plasma membrane monoamine transporter. Mol Pharmacol 68:1397–1407
9. Amphoux A, Vialou V, Drescher E, Bruss M, Mannoury La Cour C, Rochat C et al (2006) Differential pharmacological in vitro properties of organic cation transporters and regional distribution in rat brain. Neuropharmacology 50:941–952
10. Vialou V, Balasse L, Dumas S, Giros B, Gautron S (2007) Neurochemical characterization of pathways expressing plasma membrane monoamine transporter in the rat brain. Neuroscience 144:616–622
11. Vialou V, Amphoux A, Zwart R, Giros B, Gautron S (2004) Organic cation transporter 3 (Slc22a3) is implicated in salt-intake regulation. J Neurosci 24:2846–2851
12. Vialou V, Balasse L, Callebert J, Launay JM, Giros B, Gautron S (2008) Altered aminergic neurotransmission in the brain of organic cation transporter 3-deficient mice. J Neurochem 106:1471–1482
13. Cui M, Aras R, Christian WV, Rappold PM, Hatwar M, Panza J et al (2009) The organic cation transporter-3 is a pivotal modulator of neurodegeneration in the nigrostriatal dopaminergic pathway. Proc Natl Acad Sci U S A 106:8043–8048
14. Gasser PJ, Orchinik M, Raju I, Lowry CA (2009) Distribution of organic cation transporter 3, a corticosterone-sensitive monoamine transporter, in the rat brain. J Comp Neurol 512:529–555
15. Bacq A, Balasse L, Biala G, Guiard B, Gardier AM, Schinkel A et al (2012) Organic cation transporter 2 controls brain norepinephrine and serotonin clearance and antidepressant response. Mol Psychiatry 17:926–939
16. Dahlin A, Xia L, Kong W, Hevner R, Wang J (2007) Expression and immunolocalization of the plasma membrane monoamine transporter in the brain. Neuroscience 146:1193–1211

Chapter 7

Analysis of Neurotransmitter Transporter Posttranslational Modifications: Phosphorylation and Palmitoylation

James D. Foster, Danielle E. Rastedt, Sathyavathi ChallaSivaKanaka, and Roxanne A. Vaughan

Abstract

Posttranslational modification is a key mechanism by which proteins are regulated. Phosphorylation is one of the most prevalent mechanisms of regulation and fatty acid modification of proteins is becoming more widely appreciated as a regulatory mechanism for many proteins. The dopamine transporter (DAT) is an integral plasma membrane protein that mediates reuptake of dopamine from the extraneuronal space and is subject to several posttranslational modifications including phosphorylation and palmitoylation. DAT belongs to a family of closely related neurotransmitter transporters that includes carriers for serotonin and norepinephrine, and dysregulation of these transporters is hypothesized to contribute to many neuropsychiatric disorders such as depression, schizophrenia, attention-deficit hyperactivity disorder, and Parkinson disease. Neurotransmitter transporter phosphorylation and palmitoylation can be studied using various methods including metabolic labeling, phosphoamino acid analysis, peptide mapping, and acyl-biotinyl exchange. In this chapter, we discuss these methods that are employed in our laboratory for studying these modifications on DAT.

Key words Metabolic labeling, Phosphoamino acid analysis, phosphopeptide mapping, Phosphospecific antibodies

1 Introduction

The SLC6 family of Na^+/Cl^--dependent transporters includes carriers for dopamine (DA), serotonin (5-HT), norepinephrine (NE), gamma-aminobutyric acid, taurine, creatine, as well as several amino acids [1]. The monoamines DA, 5-HT, and NE control many functions including movement, mood, motivation, cognition, and reward, and their respective transporters DAT, SERT, and NET are responsible for the spatial and temporal control of extraneuronal neurotransmitter levels by driving the reuptake of extracellular transmitter into presynaptic neurons. Disruption in brain monoamine levels has been associated with several diseases such as depression, bipolar disorder, anxiety, attention-deficit

Heinz Bönisch and Harald H. Sitte (eds.), *Neurotransmitter Transporters: Investigative Methods*, Neuromethods, vol. 118, DOI 10.1007/978-1-4939-3765-3_7,

hyperactivity disorder, obsessive–compulsive disorder, Parkinson disease, and Tourette syndrome. In that light, monoamine transporters are the targets of several therapeutic drugs used to alter extraneuronal levels of these neurotransmitters in the management of these disorders. In addition, these transporters are the target of psychostimulant drugs including cocaine and amphetamines (amphetamine, methamphetamine, and 3,4,-methylenedioxy-*N*-methamphetamine) that elevate extraneuronal levels of monoamines.

Although monoamine neurotransmitter transporters were once thought to be static in terms of their activity and membrane levels, it is now clear that these transporters are regulated, providing neurons the ability to modify transport velocities in response to physiological stimuli. Transporter function is modulated through posttranslational modifications and interaction with binding partners that alter transporter kinetics, trafficking, and cell surface levels. A growing body of evidence suggests that disruption of these regulatory processes may be factors in related disease etiologies making the identification and characterization of these posttranslational modifications a high priority [2].

Our laboratory has examined the regulation of DAT by phosphorylation for several years [3–5] and more recently we have begun studying DAT palmitoylation and its effect on transporter activity, trafficking, and degradation [6]. We have employed strategic combinations of several methods to identify phosphorylation and palmitoylation sites in the transporter. These techniques include metabolic radiolabeling with $[^{32}P]H_3PO_4$ or $[^3H]$palmitic acid, phosphoamino acid analysis, 2D phosphopeptide mapping, in vitro N- and C-terminal peptide phosphorylation analysis, DAT site-specific mutagenesis, acyl-biotinyl exchange (ABE), and the generation of phosphospecific antibodies. The challenges in studying these DAT modifications lie in the fact that multiple phosphorylation and palmitoylation sites are present that are acted upon by multiple enzymes. These characteristics in addition to differences in expression levels of WT and mutant DAT proteins make it difficult to compare posttranslational modifications without very careful control of these variables.

2 Equipment and Materials

Although the techniques described here utilize many materials and equipment commonly found in research laboratories, we have listed key reagents, materials, and recipes for buffers used that we find give consistent and successful results.

3 Procedures/Methods

3.1 Phosphorylation

3.1.1 $[^{32}P]H_3PO_4$ Metabolic Labeling

The incorporation of radiolabeled H_3PO_4 into proteins within cells and tissues requires that these materials are metabolically viable so that the $[^{32}P]PO_4$ that enters the cells can be enzymatically incorporated into the intracellular ATP pool. Subsequently, this labeled pool can be utilized by active kinases to phosphorylate their protein substrates to various extents depending on cellular conditions. Phosphorylated DATs are prepared from rat striatal slices or cells in culture by metabolically labeling with $[^{32}P]H_3PO_4$ using procedures adapted from Halpain et al. [7]. $[^{32}P]$ is a high-energy β emitter making it potentially harmful if not used safely. It is essential that radioisotopes be used in accordance with institutional use and disposal polices and that proper safety precautions/equipment such as acrylic shielding and storage boxes, survey meters, gloves, eye protection and personal dosimeters are used to limit exposure and contain the radioactive material.

Tissue—The process is initiated by the dissection, isolation, and weighing of the tissue of interest (e.g. rat striatum) followed by slicing the tissue into 350 μm slices using a McIlwain tissue chopper. Equivalent amounts of tissue (4–8 slices) are placed into wells of a 12-well culture plate containing 0.6 ml of oxygenated Krebs-bicarbonate buffer (KBB, Table 1). Slices are preincubated for 30 min at 30 °C, with shaking at 105 rpm, followed by exchange with fresh KBB containing 1 mCi/ml $[^{32}P]H_3PO_4$ and continued incubation with shaking at 30 °C for 90–120 min. Oxygen (95 % O_2:5 % CO_2) is gently blown across the top of the plate during the incubation and treatments with test compounds or vehicle are added for the final 30 min. Test compounds (e.g. okadaic acid (OA) and 1-oleoyl-2-acetyl-*sn*-glycerol (OAG)) are dissolved at high concentrations in dimethyl sulfoxide (DMSO) followed by dilution in the incubation mixture so that a final DMSO concentration of 0.1 %, which by itself has no effect on the phosphorylation state of DAT, is not exceeded. At the end of the treatment, tissue slices are transferred to a microcentrifuge tube and centrifuged at 800 × *g* for 4 min at 4 °C. The supernatant fraction is removed by aspiration and 1 ml ice-cold KBB is added to resuspend and wash the slices. The samples are centrifuged again at 800 × *g* for 4 min at 4 °C followed by removal of the supernatant fraction and addition of 1 ml ice-cold KBB. The tissue is disrupted by four passages through a 26-gauge needle followed by centrifugation at 10,000 × *g* for 10 min at 4 °C, and the supernatant fraction is removed from the sedimented membranes. For phosphoamino acid and phosphopeptide mapping procedures, membranes are solubilized with 0.5 % SDS Laemmlie sample buffer (Table 2) at 20 mg/ml original wet tissue weight.

Table 1
Key reagents and equipment used in experimental procedures

Key reagents/equipment	Manufacturer/source
[^{32}P] H_3PO_4 (400–800 mCi/mL)	MP Biomedicals
[9-10-^{3}H] palmitic acid (1 mCi/mL; 30–60 Ci/mmol)	Moravek
Phosphoamino acid standards	Sigma-Aldrich
Dulbecco's Modified Eagle Medium (DMEM) (Phosphate-free form)	(Gibco® Life Technologies)
Potter-Elvehjem homogenizer with 0.1–0.15 mm teflon pestle clearance	VWR or Fisher Scientific
Dounce homigenizer with tight (0.063 mm) pestle clearance	VWR or Fisher Scientific
Microcrystalin cellulose thin-layer glass plates	EMD Chemicals via VWR or Fisher Scientific
McIllwain Tissue Chopper	Stoeling Co. via VWR or Fisher Scientific
Hunter thin-layer electrophoresis unit	C.B.S. Scientific via VWR or Fisher Scientific
Fluoro-Hance	Research Products International
MMTS	Pierce-Thermo Scientific
HPDP-biotin	Pierce-Thermo Scientific
NeutrAvidin agarose high-capacity resin	Pierce-Thermo Scientific
Anti-dopamine transporter antibody (MAB16, clone V16-23-23)	EDM Millipore, Novus Biologicals, Abcam, and ImmuQuest

Table 2
Recipes for buffers used in 32P metabolic labeling

Buffer	Recipe
Krebs-bicarbonate buffer (KBB)	25 mM $NaHCO_3$, 125 mM NaCl, 5 mM KCl, 1.5 mM $CaCl_2$, 5 mM $MgSO_4$, and 10 mM glucose, pH 7.3 (oxygenate by infusing with 95% O_2/5% CO_2 for 30–45 min on ice)
0.5% SDS Laemmli sample buffer	60 mM Tris, 0.5% SDS, 10% glycerol, 100 mM DTT, 2% 2-mercaptoethanol at pH 6.8
Buffer B	0.25 M sucrose, 10 mM triethanolamine, 10 mM acetic acid at pH 7.8
Lysis buffer	10 mM triethanolamine acetate, 150 mM NaCl, 0.1% Triton X-100, 15% sucrose, 100 mM dithiothreital, protease inhibitor at pH 7.8
Immunoprecipitation buffer (IP)	50 mM Tris–HCl, 0.1% Triton X-100 at pH 8.0
Laemmli sample buffer	60 mM Tris, 2% SDS, 10% glycerol, 100 mM DTT, 2% 2-mercaptoethanol at pH 6.8

Cells in culture—Cells are grown in 6-well culture plates to ~70–80% confluency. The culture medium is removed from each well by aspiration and replaced with 1 ml 37 °C phosphate-free DMEM followed by incubation at 37 °C in a 5% CO_2-gassed incubator. After 1 h the medium is removed by aspiration and replaced with 1 ml fresh phosphate-free DMEM at 37 °C supplemented with [^{32}P]H_3PO_4 (0.5 mCi/ml). The culture plates are placed inside an acrylic box with the lid slightly ajar for air exchange and positioned within a 5% CO_2-gassed incubator at 37 °C. After 2 h the cells can be treated with test compounds or vehicle, which are added in small volumes (1–10 μl), for the desired treatment time. Then the cells are placed on ice and the labeling medium is removed by aspiration followed by addition of 0.5 ml ice-cold buffer B (Table 2) and aspiration of the wash buffer. The labeled cells are washed in this manner two more times taking care not to dislodge them from the well surface. After the final wash, 0.5 ml buffer B is added to each well and the cells are harvested by scraping and gathering them on one side of the well followed by transfer of the cell mass and buffer to a labeled 1.5 ml microcentrifuge tube. The cells are pelleted by centrifugation at 2000×*g* for 5 min at 4 °C and the supernatant fraction is removed by aspiration. The cell mass from a single well is gently resuspended in 750 μl of lysis buffer (Table 2) followed by end-over-end mixing for 10 min at 4 °C and subsequent centrifugation at 4000×*g* for 5 min at 4 °C. The supernatant fraction is transferred to a new 1.5 ml microcentrifuge tube being careful not to transfer any debris and 200 μl of 2% SDS is added followed by thorough mixing and centrifugation at 20,000×*g* for 30 min at 4 °C. The ^{32}P-labeled supernatant is then transferred to a new microcentrifuge tube and total DAT protein levels are assessed by immunoblot blot analysis, which is essential to ensure that equal amounts of DAT protein are subjected to immunoprecipitation so that valid comparisons are made between experimental treatments.

Immunoprecipitation—Equal amounts of DAT protein are immunoprecipitated from either cell or tissue lysates by diluting lysate samples (~200 μl) 1:5 with immunoprecipitation (IP) buffer (Table 2) to reduce the SDS concentration to 0.1%. IP is performed by adding 50 μl of a 50% slurry of protein A or G agarose resin along with 0.1–0.4 μg of the antibody to the diluted lysate. Alternatively, the antibody can be covalently cross-linked to the resin in advance using dimethyl pimelimidate [8] and stored for subsequent immunoprecipitation procedures. After end-over-end mixing overnight at 4 °C, the IP resin is washed three times by repeated centrifugation and resuspension using IP buffer containing 0.1% SDS. DATs are eluted from the resin with 32 μl Laemmli sample buffer (Table 2) and subjected to SDS-PAGE followed by drying of the gel on thick filter paper and exposure to X-ray film in the presence of a quality intensifying screen at −70 to 80 °C for 1–4 days.

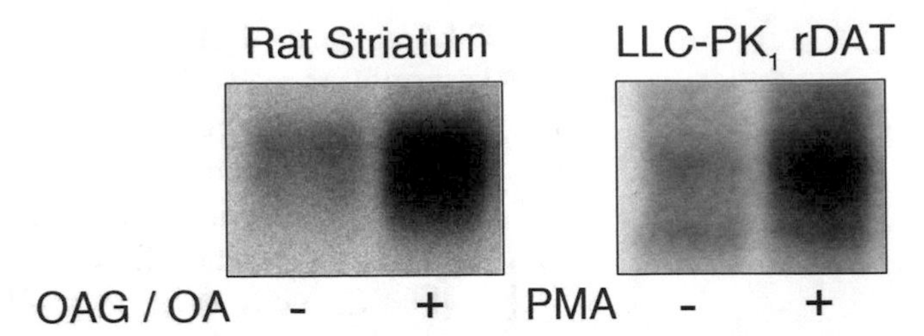

Fig. 1 Autoradiogram of immunoprecipitated DAT after ^{32}P metabolic labeling. Equal amounts of sample from rat striatal tissue or LLC-PK$_1$ cells expressing rat DAT were subjected to immunoprecipitation, electrophoresis, and autoradiography. Where indicated, tissue or cells were treated with vehicle (–) or 1 μM OAG/OA or PMA (+), respectively, for 30 min prior to solubilization and subsequent immunoprecipitation

A typical experimental result is shown in Fig. 1. Phosphate incorporation into DAT is easily detected in untreated tissue or cultured cell expression systems demonstrating basal levels of phosphorylation, and stimulated phosphorylation is seen in the presence of the protein kinase C (PKC) activators OAG or phorbol-12-myristate 13-acetate (PMA) and the PP1/PP2A phosphatase inhibitor OA. In tissue slices, maximal stimulation of DAT phosphorylation is achieved in the presence of OAG and OA together (Fig. 1), although OA treatment alone produces near maximal stimulation of DAT phosphorylation. In heterologous cell systems, treatment with PMA stimulates maximum DAT phosphorylation and is routinely used as a positive control for stimulation of PKC-dependent phosphorylation.

Notes

1. [^{32}P] is a high-energy β emitter and all procedures using this isotope should be performed behind 3/8 to ½ in. acrylic plastic shielding. Radiolabeled samples should be transported and stored in acrylic shielded containers or blocks.
2. All aspirations of [^{32}P]-containing liquids should be performed so that the liquid is pulled into a labeled waste flask housed within an acrylic plastic shielded container.
3. Incubations with tissue are performed in a sealed plastic container fitted with a hose barb for delivery of O_2 and a removable lid with a small (5 mm) diameter vent hole.
4. Lysates are adjusted to contain equal amounts of total DAT by dilution with buffer identical in composition to the lysate (750 μl lysis buffer (Table 2) plus 200 μl 2% SDS).

3.1.2 Phosphoamino Acid Analysis

The first and most straightforward line of analysis in studying phosphoproteins is to determine which of the three phosphoamino acids, phospho-serine (pS), -threonine (pT), or -tyrosine (pY), is present in the protein. This information in combination with predicted kinase consensus sequence information can provide insight into potential sites of phosphorylation within a protein and the steps that will need to be taken to investigate these sites. Phosphoamino acid analysis is performed using the method of Boyle et al. [9]. Cells or tissue samples are $[^{32}P]H_3PO_4$-labeled, treated with vehicle or test compounds, and immunoprecipitated as described above. Following SDS-PAGE, the gels are dried on thick filter paper and exposed to X-ray film in the presence of a quality intensifying screen at −70 to 80 °C for 1–2 days to reveal the radiolabeled DATs. Fluorescent markers are placed on the paper backing to allow alignment of the developed film to the gel for excision of the region containing DAT. Gel pieces from 8 to 9 lanes are rehydrated with gel running buffer (Table 3), the paper backing is removed, and the rehydrated gel is minced into small pieces. DAT is electroeluted from the pieces for 4–5 h in running buffer (Table 3) according to the manufacturer's instructions. The electroeluted DAT sample (~800 μl) is dialyzed against deionized distilled water overnight and then is acetone-precipitated to remove SDS and buffer salts. This is achieved by the addition of 4 volumes of acetone to the sample followed by centrifugation at 14,000×*g* for 10 min at room temperature (RT). The supernatant fraction is removed by aspiration and the DAT protein in the pellet is

Table 3
Recipes for buffers used in phosphoamino acid analysis and phosphopeptide mapping

Buffer/reagent	Recipe
Gel running buffer	25 mM Tris, 192 mM glycine, 0.1 % SDS ~pH 8.3; do not adjust
Ammonium bicarbonate	100 mM ~pH 8.0; do not adjust
HCl	5.7 M
Performic acid	900 μl (~98 %) formic acid plus 100 μl 30 % hydrogen peroxide; 60 min at room temperature; then place on ice
pH 1.9 $buffer^{electrophoresis}$	7.8 % acetic acid, 2.5 % formic acid pH 1.8–1.9; do not adjust
pH 3.5 $buffer^{electrophoresis}$	0.5 % pyridine, 5 % acetic acid ~pH 3.5; do not adjust
pH 3.5 $buffer^{chromatography}$	25 % pyridine, 7.5 % acetic acid, 37.5 % *n*-butanol ~pH 3.5; do not adjust
Ninhydrin	0.25 % in ethanol (spray applicator)
Phosphoamino acid standards	1 mg/ml of each phosphoserine, phosphotyrosine, and phosphothreonine in deionized water

subjected to partial acid hydrolysis in 50–100 μl 5.7 M HCl for 1 h at 110 °C. The hydrolyzed sample is lyophilized to dryness in a centrifugal evaporator designed for acidic and basic solutions and the dried hydrolysate is dissolved in ~10 μl of pH 1.9 buffer (Table 3) containing phosphoamino acid standards (30 parts buffer and 1 part standard, Table 3). The sample is briefly centrifuged to remove any particulates and spotted onto a microcrystalline cellulose thin-layer plate (Table 1) in increments of ~0.5 μl with drying of the spot between applications.

Using a Hunter thin-layer electrophoresis unit (Table 1) [9] the samples are separated at 1.5 kV for 35 min at pH 1.9 in the first dimension, and at 1.3 kV for 20 min at pH 3.5 in the second dimension. Standards are visualized by spraying the plate with ninhydrin (mauve/magenta colored staining after 5 min incubation at 65 °C) and the dried plates are subjected to autoradiography for 10–14 days. Self-adhering fluorescent markers are applied to the plate prior to autoradiography so that the resulting autoradiogram can be aligned with the plate to overlay the radiolabeled spots with the ninhydrin-stained phosphoamino acid standards. A typical phosphoamino acid analysis is shown in Fig. 2. Negatively charged amino acids migrate to the left in each dimension so that the phosphoamino acids pS, pT, and pY are found in the upper left quadrant of the plate after two-dimensional electrophoresis. Because this is a partial acid hydrolysis procedure, ^{32}P-labeled peptides will also be observed but will have less mobility and remain closer to the origin than the free amino acids. Rat striatal DAT consistently exhibits a pattern of ~90% pS and ~10% pT. In numerous attempts we have been unable to detect pY, which is labile under acidic conditions and may require alternative methods for detection.

Notes

1. Detailed instructions on performing 2D thin-layer electrophoresis can be found in [9] along with further details regarding phosphoamino acid analysis.
2. In the process of spotting the hydrolyzed sample onto the thin-layer plates, it is essential that no particulates are transferred to the plate which will cause streaking and suboptimal sample separation.

3.1.3 Phosphopeptide Mapping

Phosphopeptide mapping is the combination of ^{32}P metabolic labeling, proteolytic digestion, and separation of the digestion products to produce a unique phosphopeptide fingerprint. The peptide map is dependent on the proteolytic enzyme employed and may provide information related to the location of phosphorylation sites in the protein of interest. The map also provides information on phosphorylation states or sites of phosphorylation

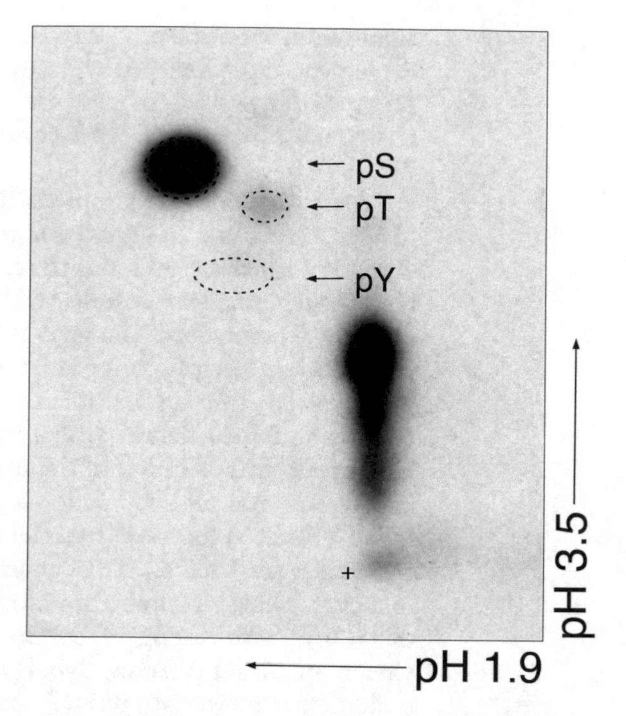

Fig. 2 Phosphoamino acid analysis of rat DAT expressed in LLC-PK_1 cells. $^{32}PO_4$-labeled cells were treated with 1 μM PMA for 30 min prior to purification by immunoprecipitation and gel electrophoresis. Autoradiographs displayed the phosphorylation pattern demonstrated in Fig. 1 (*far right lane*). DAT bands were excised, eluted, and subjected to phosphoamino acid analysis and autoradiography. Phosphoamino acid standards were visualized with ninhydrin (*dotted circles*). *pS* phosphoserine, *pT* phosphothreonine, *pY* phosphotyrosine

that may occur upon changes in cellular conditions. A number of different proteases or chemicals with unique cleavage characteristics can be used to produce maps that aid in identifying specific phosphorylation sites within a protein.

The procedure begins with ^{32}P metabolic labeling, gel purification, electroelution, and acetone precipitation as described above for phosphoamino acid analysis. Once the target protein is labeled and isolated, the precipitate is dissolved in 50–100 μl of ice-cold performic acid (Table 3) and incubated on ice for 60 min. This step oxidizes the methionine and cysteine residues present in proteins and is necessary for disulfide bond cleavage. This simplifies the map by eliminating fragments held together by disulfide bonds and can be omitted if these residues are not present in the peptides of interest. After oxidation, 400 μl of purified water is added to dilute the performic acid and the sample is frozen on dry ice followed by lyophilization in a centrifugal evaporator designed for acidic and basic solutions. The resulting dried protein pellet is prepared for proteolytic cleavage by resuspension in 50 μl of ammonium bicarbonate, pH 8.0–8.3 or another volatile buffer. Proteolytic digestions are typically carried out in the presence of 10 μg of the proteolytic enzyme for 5 h at 37 °C followed by a second dose of enzyme and continued incubation overnight at 37 °C.

After digestion the sample is lyophilized to dryness in a centrifugal evaporator designed for acidic and basic solutions. The digest is washed several times to remove the volatile buffer by adding 400 μl purified water followed by lyophilization to dryness in a centrifugal evaporator. The final 400 μl resuspension is centrifuged for 2 min in a microfuge at maximum speed and the supernatant is transferred to a fresh tube. It is important not to transfer any contaminating salt or debris, which will cause the peptides to streak and reduce the resolution of the map. This final solution is lyophilized to dryness and the pellet is dissolved in ~10 μl of pH 1.9 buffer (Table 3) followed by brief centrifugation to sediment any remaining particulates. The labeled digest is then spotted on a microcrystalline cellulose thin-layer plate (Table 1) in increments of ~0.5 μl with drying of the spot between applications with a stream of air. A marker dye (green) composed of 5 mg/ml ε-dinitrophenyl-lycine (yellow) and 1 mg/ml xylene cyanol FF (blue) can be applied near the top center of the plate to monitor the progress of the electrophoresis and confirm that sufficient migration has occurred as the dye separates into its yellow and green components.

The plate is subjected to thin-layer electrophoresis in the first dimension using pH 1.9 buffer for 35 min at 1500 V in a Hunter thin-layer electrophoresis unit (Table 1) followed by air drying. The plate is then subjected to ascending chromatography in the second dimension using pH 3.5 chromatography buffer (Table 3). After the buffer has ascended within ~1 cm of the top of the plate, the plate is removed from the chromatography chamber and air-dried. Self-adhering fluorescent makers are applied to the plate so that the subsequent autoradiogram can be aligned to the plate and establish the origin on the exposed film. The plate is exposed to x-ray film in the presence of a quality intensifying screen at −70 to 80 °C for 10–14 days. In the first dimension, peptides with a net positive charge will migrate to the right and those with a net negative charge will move to the left, with the distance of migration from the origin dependent on the peptide mass/charge ratio. A typical phosphopeptide mapping experiment is shown in Fig. 3 showing several DAT phosphopeptides (spots 2–9). Significant OAG/OA stimulation of phosphorylation is seen in spots 7–9 indicating that phosphorylation sites in these peptides are utilized in the untreated and stimulated state. The large spot of radioactivity on the upper left side in each map (spot 1) is free $^{32}PO_4$, which is released from the protein during the procedure.

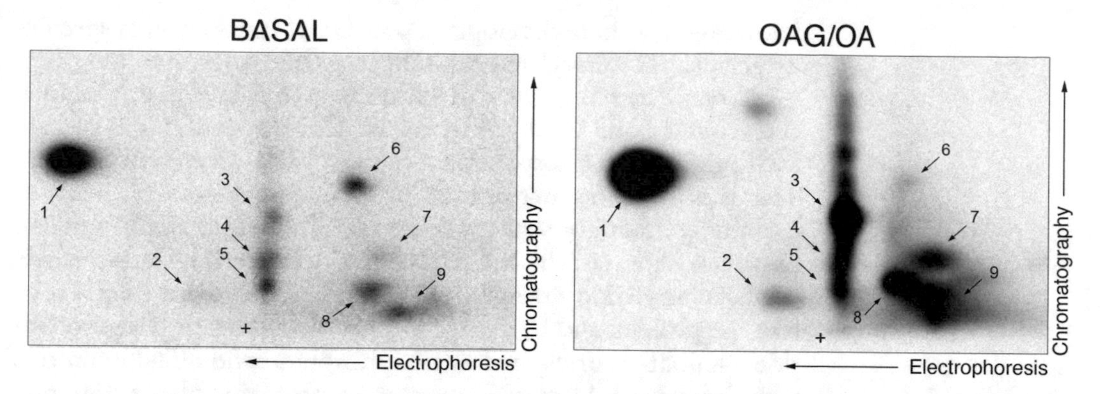

Fig. 3 Phosphopeptide mapping analysis of rat striatal DAT. $^{32}PO_4$-labeled striatal tissue was treated with vehicle (Basal) or 1 μM OAG/OA for 30 min prior to purification by immunoprecipitation and gel electrophoresis. Autoradiographs displayed the phosphorylation pattern demonstrated in Fig. 1 (*left panel*). DAT bands were excised, eluted, and subjected to proteolytic digestion with thermolysin. Samples were subjected to 2D electrophoresis/chromatography and autoradiography. Peptide spots were numbered for comparison between treatments

Notes

1. Detailed instructions on performing 2D thin-layer electrophoresis/chromatography can be found in [9] along with further details regarding phosphopeptide mapping.
2. In the process of spotting the hydrolyzed sample onto the thin-layer plates it is essential that no particulates are transferred to the plate which will cause streaking and suboptimal separation of the peptides.

3.1.4 Phosphospecific Antibodies

Phosphoamino acid analysis and phosphopeptide mapping are very useful in determining protein phosphorylation sites but they are labor-intensive and require the use of high levels of radioactivity. In addition, many proteins, such as DAT, are phosphorylated at multiple residues making it difficult to measure phosphorylation at a specific site by metabolic labeling. An alternative method for detecting and quantifying phosphorylation sites that circumvents these issues is the use of phosphospecific antibodies. If good evidence exists for the location of a particular phosphorylation site, antibodies specific for that site within the protein of interest can be developed to detect and quantify its phosphorylation properties.

Two sites of phosphorylation have been identified for DATs [4, 5]. The first, S7, is found within a distal N-terminal serine cluster that undergoes increased phosphorylation in response to activation of PKC and by amphetamine (AMPH) and methamphetamine (METH) treatments [10, 11]. The second is a proline-directed phosphorylation site in the membrane proximal region of the N-terminus at rat/mouse T53 and human DAT S53 [4]. With this knowledge, we have developed a polyclonal antibody with specificity against phosphorylated T53 DAT (pT53 Ab).

When generating such an antibody, it is critical that it be fully characterized and validated to ensure that the antibody is specific for the intended phosphorylation site in the protein of interest. This is achieved by performing proper positive controls, such as examining immunoreactivity under conditions known to stimulate phosphorylation of the site, and negative controls such as analysis of phosphorylation site mutants and tissue or cells that do not contain the protein, and by antiserum preabsorbed with phosphopeptide but not the corresponding nonphosphopeptide. These controls will also demonstrate any nonspecific reactivity of the antibody with other proteins.

To detect phosphorylation of T53, lysates of tissue or cells are subjected to SDS-PAGE and the proteins are transferred to PVDF membrane. The membrane is blocked for at least 1 h in blocking buffer (Table 4) at RT followed by incubation overnight at 4 °C with pT53 Ab (primary antibody) diluted 1:1000 with blocking buffer. The membrane is then washed five times with wash buffer (Table 4) followed by incubation of the membrane for 1 h at RT with alkaline phosphatase-conjugated anti-rabbit IgG Ab (secondary antibody) diluted 1:5000 with blocking buffer. The membrane is again washed five times for 5 min with wash buffer and incubated with 3 ml of alkaline phosphatase chemiluminescent substrate for 5 min at RT. The wet membrane is inserted between two transparent plastic sheets and immunoreactive bands are visualized with a Bio-Rad Image Doc System.

A typical analysis using pT53 Ab to detect DAT phosphorylation is shown in Fig. 4. pT53 Ab produces robust immunoreactivity with lysates from cells or striatal tissue expressing rat DAT (rDAT) with little immunostaining in LLC-PK_1 cells or rat brain cerebellum, which do not express DAT. Treatment of cells with the PKC activator, PMA, or with the protein phosphatase inhibitor, OA, results in significant elevation of pT53 immunoreactivity (Fig. 4b).

In addition to recognizing rDAT, pT53 Ab also recognizes mouse DAT (mDAT), which both contain T53-P54 in their primary sequence, but does not react with human DAT, which

Table 4
Recipes for buffers used in phosphospecific immunoblotting analysis

Buffer	Recipe
Blocking buffer	3% BSA in PBS at pH 7.4
Wash buffer	PBS at pH 7.4 containing 0.1% TWEEN

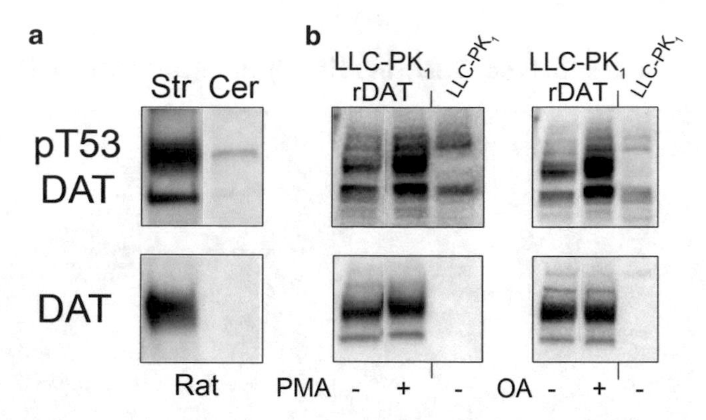

Fig. 4 (**a**) Rat DAT immunoblot analysis with pT53 phosphospecific DAT antibody. Lysates from rat striatum (*Str*) and cerebellum (*Cer*) or (**b**) lysates from LLC-PK$_1$ or rDAT-LLCPK$_1$ cells were immunoblotted with pT53 Ab (*upper panels*) or MAb16 (*lower panels*). Where indicated cells were treated with 1 μM PMA or OA for 30 min at 37 °C prior to solubilization

contains a serine rather than a threonine at this position [4]. Our results also show a level of cross-reactivity with unknown proteins in the cerebellum and LLC-PK$_1$ cells that do not express DAT, demonstrating the necessity for performing proper controls (see Fig. 4). The pT53 Ab can also be used for immunoprecipitation followed by immunoblotting the precipitated DATs with a monoclonal antibody (MAb16, Table 1) specific for total DATs [12].

3.2 Palmitoylation

3.2.1 Metabolic Labeling

One of the first techniques developed and still widely used to study protein acylation is metabolic labeling with radiolabeled fatty acids [13]. Metabolic labeling with [^{3}H]palmitic acid can be performed in cultured cells lines or in tissue samples such as rat striatal synaptosomes. As with metabolic phosphorylation, metabolic palmitoylation requires viable tissue or cells so that the fatty acid can be translocated into the cytoplasm, metabolized into its active CoA form, and ultimately transferred enzymatically to the protein of interest. Although [^{3}H] is a low-energy β emitter that does not pose an external hazard, this isotope must still be used in accordance with institutional guidelines and waste disposal polices. Because [^{3}H] cannot be detected with a survey meter, it is essential to use bench pads to absorb and contain spills and monitor the work area using swipe tests. In addition, this isotope does pose an internal hazard so it is important to use personal protection such as gloves and eyewear.

Tissue—The process is initiated by the dissection, isolation, and weighing of rat brain striatal tissue followed by preparation of synaptosomes [14]. Briefly, striata are rapidly dissected, weighed, and placed in ice-cold sucrose phosphate (SP) buffer (Table 5). Tissue (60–80 mg original wet weight) is homogenized in 2 ml of

Table 5
Recipes for reagents and buffers used in [^{3}H] palmitate metabolic labeling

Buffer	Recipe
Sucrose-Phosphate (SP) buffer	0.32 M sucrose and 10 mM sodium phosphate at pH 7.4
Krebs-Ringer-HEPES (KRH) buffer	25 mM HEPES, 125 mM NaCl, 4.8 mM KCl, 1.2 mM KH_2PO_4, 1.3 mM $CaCl_2$, 1.2 mM $MgSO_4$, 5.6 mM glucose at pH 7.4
Radioimmunoprecipitation assay (RIPA) buffer	50 mM Tris–HCl, 150 mM NaCl, 2 mM EDTA, 0.1% SDS, 1% Triton X-100 at pH 7.4
Immunoprecipitation (IP) buffer	25 mM Tris–HCl, 0.1% Triton X-100 at pH 8.0

ice-cold SP buffer with 15 strokes in a glass/Teflon homogenizer and the resulting homogenate is transferred to 15 ml tube, diluted with 8 ml of SP buffer and centrifuged at 3000 × *g* for 3 min at 4 °C. The resulting supernatant fraction is centrifuged again at 17,000 × *g* for 12 min to pellet the synaptosomal fraction. The resulting synaptosomal pellet is resuspended in labeling medium (60–80 mg original wet weight/ml) for metabolic labeling with 40 μM [9,10-^{3}H]palmitic acid (2 mCi/ml) at 30 °C for 90 min with end-over-end mixing. This is a higher level of [^{3}H]palmitate than is needed for cells because synaptosomes are viable for only a short period of time at 30 °C. The labeling medium is prepared by drying down 2 ml of [9,10-^{3}H]palmitic acid (1 mCi/ml in ethanol) in a centrifugal evaporator to ~20 μl and adding 1 mL HEPES buffered DMEM. Following labeling, the synaptosomes are pelleted by centrifugation at 17,000 × *g* for 12 min at 4 °C. The supernatant fraction is removed by aspiration and the pellet is washed with 1 ml SP buffer and centrifuged again at 17,000 × *g* for 12 min at 4 °C. The supernatant fraction is removed and the final protein pellet is solubilized in 1 ml of RIPA buffer (Table 5). Labeled DATs are isolated by immunoprecipitation after centrifugation of the lysate at 20,000 × *g* for 30 min at 4 °C to remove insoluble material (see *Immunoprecipitation* above).

Cells in culture—Cells are grown to 80–90% confluency in 6- or 12-well plates and incubated for 18 h at 37 °C with culture medium containing 10 μM [9,10-^{3}H] palmitic acid (0.5 mCi/ml). The labeling medium is prepared by drying down 2 ml of [9,10-^{3}H] palmitic acid (1 mCi/ml in ethanol) in a centrifugal evaporator to ~100 μl. This solution is then added to 4 ml of cell culture medium and mixed thoroughly. The medium also contains 1 mM sodium pyruvate, which acts as a source of acetyl-CoA and inhibits palmitate metabolism via fatty acid β-oxidation. After labeling, cells are washed twice with 1–2 ml of Krebs-Ringers-HEPES buffer (Table 5) and lysed by the addition of 800 μl ice-cold RIPA buffer.

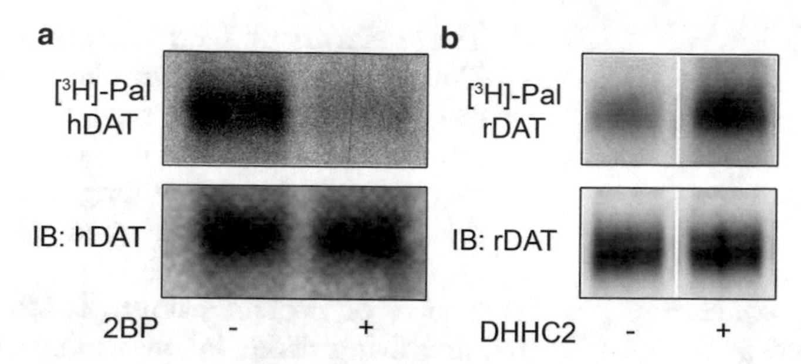

Fig. 5 Detection of DAT palmitoylation analyzed by [^{3}H]palmitic acid labeling. (**a**) 6xHis hDAT-HEK 293 cells were labeled with 0.5 mCi/ml (10 μM) [^{3}H]palmitate in the presence (+) or absence (–) of 15 μM 2BP for 18 h at 37 °C. (**b**) LLCPK$_1$ cells stably expressing rDAT were transiently transfected DHHC2 cDNA (+) or empty vector (–). After 4 h the cells were labeled with 10 μM [^{3}H]palmitate for 18 h at 37 °C followed by solubilization. Equal amounts of DAT were immunoprecipitated and analyzed by SDS-PAGE/fluorography ([^{3}H]pal) or immunoblotting (IB)

The lysate is transferred to microfuge tubes and centrifuged at 20,000 × *g* for 30 min to remove nuclei and cell debris.

Labeled DATs are immunoprecipitated from the cleared lysate as described above (see *Immunoprecipitation*) and subjected to SDS-PAGE. The gel is soaked in a fluorographic reagent (Fluoro-Hance) for 30 min, dried, and exposed to preflashed X-ray film for 1–6 weeks. Flashing the film prior to exposure with [^{3}H] allows for the detection of weak signals that would normally not be observed. Band labeling intensities are quantified using imaging software such as Quantity One® (Bio-Rad). Parallel aliquots of the lysate are immunoblotted to determine total DAT levels and [^{3}H]palmitate DAT labeling is normalized to the DAT protein levels.

Palmitoyl acyl transferases (PAT or DHHC) are enzymes that catalyze protein palmitoylation, and their activity can be modulated by heterologous overexpression or treatment with the irreversible PAT inhibitor 2-bromopalmitate (2BP). Modulation of PAT activity results in changes in protein palmitoylation that allows for the determination of mechanistic roles this modification may play in cellular events. A typical [^{3}H]palmitate metabolic labeling result for DAT is shown in Fig. 5. DAT labeling is decreased in the presence of 2BP (Fig. 5a) but is increased by DHHC2 co-expression (Fig. 5b).

Notes

1. [^{3}H] is a low-energy β emitter and these particles do not effectively escape the dried gel matrix. Therefore, it is necessary to use a fluorographic reagent to impregnate the gel enabling β particles to interact with the scintillant thereby producing photons that will escape the gel and expose the film.

2. The sensitivity of X-ray film can be increased by preflashing the film by pre-exposing the film to an instantaneous flash of filtered light (450–495 nm) which overcomes the nonlinear response of film to low intensities of light. It is also important that the emulsion-containing side of the film is flashed and this side of the film faces the dried gel.

3.2.2 Acyl-biotinyl exchange (ABE)

Detection of protein palmitoylation by metabolic radiolabeling requires living tissue, involves radioactivity, has lengthy film exposure times, and is relatively low in sensitivity. An alternative method for detection of protein palmitoylation is acyl-biotinyl exchange (ABE) [15] which allows for more rapid and sensitive detection of palmitoylated proteins by substituting biotinyl moieties for the thioester-linked palmitic acid and subsequent capture of the endogenous palmitoylated protein [16]. ABE permits the detection of protein palmitoylation levels at the moment of sample lysis and can be performed using frozen tissue [17]. The ABE method involves three basic steps: (1) protein denaturation and blockade of unmodified cysteine thiols, (2) cleavage of palmitoyl thioester linkages, and (3) labeling the newly exposed cysteine sulfhydryls with a thiol-specific biotinylation reagent and capture on NeutrAvidin® resin. After each step, the reactants, which will interfere with subsequent step, are separated from the protein by three sequential acetone precipitations with the pellet resuspended after each precipitation in 4SB (Table 6).

Protein denaturation and blockade of unmodified cysteine thiols—The ABE technique, modified from Wan et al. [16], begins with dividing cell or tissue lysates into equal fractions based on protein content and precipitating the protein with the addition of 4 volumes of acetone. The supernatant is removed and the pellet is solubilized in 200 μl of lysis buffer containing protease inhibitor and 20 mM methyl methanethiosulfonate (MMTS). Protein denaturation by SDS and subsequent acetone precipitation is essential to facilitate the accessibility of MMTS to all cysteine residues within the protein. Samples are incubated at RT for 1 h while rotating end-over-end, followed by acetone precipitation and resuspension of the protein pellet in lysis buffer containing 20 mM MMTS. This process is repeated, followed by resuspension of the protein pellet in lysis buffer containing 20 mM MMTS and incubation overnight at RT with continued end-over-end mixing. The following day, the excess MMTS is removed by three sequential acetone precipitations. It is essential that all residual MMTS is removed as any that remains can modify newly generated thiols in the thioester cleavage step and thus, greatly reduce the final biotinylation reaction.

Cleavage of palmitate thioester linkages—The next step in the ABE procedure involves the cleavage of thioester-linked palmitoyl moieties. The acetone-precipitated protein pellets from the

Table 6
Recipes for buffers used in acyl-biotinyl exchange

Buffer	Recipe
Lysis buffer	50 mM HEPES, 1 mM EDTA, 2% sodium dodecyl sulfate (SDS) at pH 7.0
4% SDS buffer (4SB)	4% SDS, 50 mM Tris–HCl, 5 mM EDTA at pH 7.4
Tris buffer	50 mM Tris–HCl at pH 7.4
Hydroxylamine	0.7 M NH_2OH at pH 7.4; make fresh daily

previous step are solubilized in 300 μl of 4SB (Table 6) and divided into two equal portions that are treated with either 50 mM Tris–HCl (control), pH 7.4 or 0.7 M hydroxylamine (NH_2OH), pH 7.4 for 2 h at RT with end-over-end mixing. The NH_2OH will not affect the MMTS-modified cysteines but will cleave any endogenous palmitoyl thioester groups and render the endogenously palmitoylated sites susceptible to modification in the next step. Tris treatment is a crucial control as it does not remove endogenous palmitate and these samples should contain little to no biotinylated protein at the end of the procedure. After incubation with NH_2OH or Tris–HCl, excess reagent is removed by three sequential acetone precipitations.

Labeling the newly exposed cysteine sulfhydryls with a thiol-specific biotinylation reagent and capture on NeutrAvidin resin—The final stage of the ABE process involves modification of the newly exposed sulfhydryl group using the sulfhydryl reactive reagent (*N*-(6-(biotinamido)hexyl)-3′(2′-pyridyldithio)-propionamide) (HPDP) biotin. This reagent will tag any previously palmitoylated cysteine residues with biotin and allow the modified proteins to be isolated via NeutrAvidin chromatography. For this procedure, protein pellets from the thioester cleavage step are resuspended in 240 μl 4SB, diluted with 900 μl of 50 mM Tris–HCl, pH 7.4 containing 0.5 mM HPDP-biotin, and incubated at RT for 1 h with end-over-end mixing. Unreacted HPDP-biotin is removed by three sequential acetone precipitations. It is important to remove all HPDP-biotin as it will compete with biotinylated proteins for NeutrAvidin-agarose binding. The final protein pellet is solubilized in 75 μl lysis buffer (Table 6) and a small fraction (10 μl) is set aside to determine total DAT protein content. The remaining portion is diluted 1:20 with 50 mM Tris–HCl, pH 7.4 to reduce the SDS concentration to 0.1% for NeutrAvidin affinity chromatography. Biotinylated proteins are isolated using 25 μl of a 50% slurry of NeutrAvidin agarose resin and incubation for 2 h at RT or overnight at 4 °C with end-over-end mixing. The mixture is centrifuged at 6000 × *g* for 1 min, the unbound supernatant is removed,

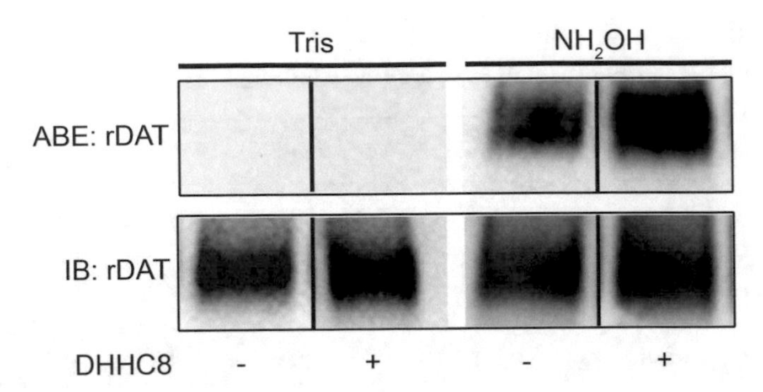

Fig. 6 Detection of DAT palmitoylation analyzed by ABE. $LLCPK_1$ cells stably expressing rDAT were transiently transfected with DHHC8 cDNA (+) or empty vector (–). After 24 h membranes were prepared followed by determination of DAT palmitoylation by ABE

and the NeutrAvidin resin is washed three times with RIPA buffer. Proteins are eluted from the resin with 32 μl Laemmli sample buffer (Table 2, 30 min RT). Total and eluted proteins are subjected to SDS-PAGE and immunoblotting with anti-rat DAT monoclonal Ab 16 (mAb16, Table 1). Band intensities are quantified using imaging software such as Quantity One® (Bio-Rad) and DAT palmitoylation levels are normalized to total DAT protein.

A typical ABE result is shown in Fig. 6. Tris-treated controls (Fig. 6, *top left*) show only a negligible DAT signal indicating successful blockage of free thiols in the first step of the procedure. Robust staining of DAT is seen in the NH_2OH-treated samples (Fig. 6, *top right*) indicating successful exchange of biotin for the endogenous palmitoyl group. Equal DAT levels are seen in Tris- and NH_2OH-treated total DAT samples demonstrating that equal amounts of DAT were loaded onto the NeutrAvidin resin (Fig. 6, *bottom panel*). Co-expression of DAT and the PAT enzyme DHHC8 results in a significant increase in DAT palmitoylation suggesting that the enzyme recognizes DAT as a substrate (Fig. 6, *top right*).

Notes

1. Additional detailed instructions for ABE can be found in [15, 16].
2. It is sometimes difficult to solubilize acetone-precipitated proteins. Protein solubility can be enhanced by briefly letting the residual acetone dissipate by air-drying before adding the solubilization buffer to the sample. In addition, the sample can be incubated with solubilization buffer in a water bath at 37 °C or an ultrasonic bath to more readily resuspend the protein.

3. In the biotinylation step, it is sometimes necessary to incubate the Tris-HPDP-biotin stock solution at 37 °C for a few min prior to use to ensure that it remains solubilized.

4 Conclusion

Monoamine transporters are proving to be highly complex proteins that are regulated by posttranslational modifications. Much remains to be determined regarding the roles these modifications play in the function, regulation, and trafficking of these transporters. The determination of the specific role each modification plays in maintaining monoamine homeostasis will be crucial toward further understanding these mechanisms and how dysregulation of these events may contribute to disease. Thus, it is necessary to identify the modified sites and understand their contribution to transporter activity and regulation. Using DAT as a model transporter, we have outlined various techniques and approaches in this chapter that are aimed at aiding researchers in accomplishing this task. Furthermore, these techniques are applicable to analyzing these modifications in other related transporters.

References

1. Pramod AB, Foster J, Carvelli L, Henry LK (2013) SLC6 transporters: structure, function, regulation, disease association and therapeutics. Mol Aspects Med 34:197–219
2. Vaughan RA, Foster JD (2013) Mechanisms of dopamine transporter regulation in normal and disease states. Trends Pharm Sci 34(9):486–496
3. Foster JD, Cervinski MA, Gorentla BK, Vaughan RA (2006) Regulation of the dopamine transporter by phosphorylation. Handb Exp Pharmacol (175):197–214
4. Foster JD, Yang JW, Moritz AE, Challasivakanaka S, Smith MA, Holy M, Wilebski K, Sitte HH, Vaughan RA (2012) Dopamine transporter phosphorylation site threonine 53 regulates substrate reuptake and amphetamine-stimulated efflux. J Biol Chem 287:29702–29712
5. Moritz AE, Foster JD, Gorentla BK, Mazei-Robison MS, Yang JW, Sitte HH, Blakely RD, Vaughan RA (2013) Phosphorylation of dopamine transporter serine 7 modulates cocaine analog binding. J Biol Chem 288:20–32
6. Foster JD, Vaughan RA (2011) Palmitoylation controls dopamine transporter kinetics, degradation, and protein kinase C-dependent regulation. J Biol Chem 286:5175–5186
7. Halpain S, Girault JA, Greengard P (1990) Activation of NMDA receptors induces dephosphorylation of DARPP-32 in rat striatal slices. Nature 343:369–372
8. Gersten DM, Marchalonis JJ (1978) A rapid, novel method for the solid-phase derivatization of IgG antibodies for immune-affinity chromatography. J Immunol Methods 24:305–309
9. Boyle WJ, van der Geer P, Hunter T (1991) Phosphopeptide mapping and phosphoamino acid analysis by two-dimensional separation on thin-layer cellulose plates. Methods Enzymol 201:110–149
10. Foster JD, Pananusorn B, Vaughan RA (2002) Dopamine transporters are phosphorylated on N-terminal serines in rat striatum. J Biol Chem 277:25178–25186
11. Cervinski MA, Foster JD, Vaughan RA (2005) Psychoactive substrates stimulate dopamine transporter phosphorylation and down-regulation by cocaine-sensitive and protein kinase C-dependent mechanisms. J Biol Chem 280:40442–40449

12. Gaffaney JD, Vaughan RA (2004) Uptake inhibitors but not substrates induce protease resistance in extracellular loop two of the dopamine transporter. Mol Pharmacol 65:692–701
13. Schlesinger MJ, Magee AI, Schmidt MF (1980) Fatty acid acylation of proteins in cultured cells. J Biol Chem 255:10021–10024
14. Dunkley PR, Jarvie PE, Heath JW, Kidd GJ, Rostas JA (1986) A rapid method for isolation of synaptosomes on Percoll gradients. Brain Res 372:115–129
15. Drisdel RC, Green WN (2004) Labeling and quantifying sites of protein palmitoylation. Biotechniques 36:276–285
16. Wan J, Roth AF, Bailey AO, Davis NG (2007) Palmitoylated proteins: purification and identification. Nat Protoc 2:1573–1584
17. Fukata Y, Fukata M (2010) Protein palmitoylation in neuronal development and synaptic plasticity. Nat Rev Neurosci 11:161–175

Chapter 8

Ensemble and Single Quantum Dot Fluorescence Methods in Neurotransmitter Transporter Research

Oleg Kovtun and Sandra J. Rosenthal

Abstract

Subcellular localization and trafficking of neurotransmitter transporter (NTT) proteins is increasingly recognized to play a critical role in transporter-mediated neurotransmitter signaling and its regulation. To fully understand the molecular mechanisms underlying transporter regulation, it is essential to be able to visualize NTTs both at the population and single-molecule levels using advanced imaging techniques. Here, we describe three fluorescence-based methods that have been successfully applied to measure spatiotemporal changes in NTT localization and to establish dynamic imaging of individual NTT molecules using the ligand-conjugated quantum dot (QD) approach. First, we discuss how to label and image membrane NTTs in live cells using QD probes in conjunction with ensemble fluorescence microscopy. Second, we present a more quantitative, flow cytometry-based approach, particularly useful for assessing transporter internalization and recycling. Third, we describe a single-molecule microscopy labeling protocol for determining the mobility of QD-bound transporters at the plasma membrane of live cells.

Key words Quantum dot, Biological labeling, Neurotransmitter transporter, Confocal fluorescence microscopy, Flow cytometry, Biotinylated ligand, Single-molecule imaging

1 Introduction

Traditional biochemical and genetic approaches have contributed the majority of the existing research knowledge on NTT structure, function, and regulation. It is now apparent that NTT function is tightly regulated through multiple posttranslational mechanisms including interactions with a plethora of kinases, receptors, and scaffolding elements [1–5]. Consequent dynamic changes in NTT subcellular localization fundamentally impact the amplitude, duration, and specificity of NTT-mediated neurotransmitter signaling. Therefore, the ability to "see" NTTs with subcellular resolution and to monitor dynamic trafficking pathways involved in NTT regulation becomes a critical tool in advancing our understanding of the molecular mechanisms underlying NTT signaling network.

Heinz Bönisch and Harald H. Sitte (eds.), *Neurotransmitter Transporters: Investigative Methods*, Neuromethods, vol. 118,
DOI 10.1007/978-1-4939-3765-3_8, © Springer Science+Business Media New York 2016

Recent advances in fluorescence-based techniques for molecular biology permitted investigation of cellular signal transduction cascades with unprecedented spatiotemporal resolution [6, 7]. Currently, there are several classes of fluorescent probes available to investigators. Among these, the most prevalent fluorophores are organic dyes, genetically encoded fluorescent proteins, and semiconductor nanocrystals, colloquially known as quantum dots (QDs) [8–15]. General properties, advantages, and drawbacks of the aforementioned fluorophores are summarized in Table 1. Our group is primarily focused on exploiting the unique photophysical properties of QDs (excellent brightness, narrow emission spectra, broad excitation spectra, and superior photostability) to study subcellular distribution and dynamic regulation of NTTs [16–24].

There are several methodological approaches to enable specific targeting of membrane proteins in live cells with the aforementioned fluorescent probes (Table 2). Most commonly, (1) a fluorescent protein (e.g., EGFP) is fused to the terminus of the target protein and expressed in the cell of interest, or (2) a fluorophore is attached to an antibody targeting the extracellular domain of the target protein. Unfortunately, limited availability of such extracellular antibodies for NTTs and lack of suitable extracellular epitopes within the NTT structure have significantly hampered fluorescence-based investigation of NTT localization and regulation, particularly in endogenous systems. To this end, we pioneered a ligand-conjugated QD-based approach that utilizes a transporter-specific organic ligand composed of (1) a high-affinity parent drug that enables recognition of specific binding sites within the

Table 1
Comparison of commonly encountered fluorescent probes

Property	Cy5	EGFP	QD655
Size	~0.5 nm; 792 Da	~5 nm; ~27 kDa	15–25 nm; >1000 kDa
Quantum yield	0.3	0.6	0.5
Molar absorption coefficient	2.5×10^5 M^{-1} cm^{-1}	5.6×10^4 M^{-1} cm^{-1}	$>2 \times 10^6$ M^{-1} cm^{-1}
Excitation/emission maxima	649/670 nm	488/509 nm	Steady increase toward UV wavelengths starting from absorption onset; emission max at 655 nm, with FWHM ~30 nm
Photostability	5–10 s	5–10 s	Minutes
Applicability to single-molecule imaging	Moderate; limited by poor photostability	Moderate; limited by poor photostability	Excellent; complicated by blinking

Table 2
Methodological approaches that enable specific targeting of neurotransmitter transporters

Approach	Advantages\disadvantages	References
Genetic fusion (XFP; hemagglutinin fusion peptide)	Pros: perfect specificity, biocompatibility, retention of the XFP-target protein construct Cons: incompatibility with endogenous expression systems, misfolding, failure to localize properly, altered activity compared to the wild-type protein	[25–31]
Antibody	Pros: excellent specificity, biocompatibility, low cytotoxicity, compatibility with endogenous expression systems Cons: lack of efficient external antibodies, large size, prone to chemical degradation	[32]
Organic ligand	Pros: targeting specificity and selectivity, binding stability, compatibility with endogenous expression systems, biological activity Cons: sophisticated organic chemistry and rigorous analytical characterization required for preparation, primary binding site occupied	[16–24, 33–36]
Covalent modification	Pros: excellent selectivity, possibility of an inert functional tag, small size Cons: potentially deleterious effects on protein structure and function, organic chemistry knowledge required	[37, 38]

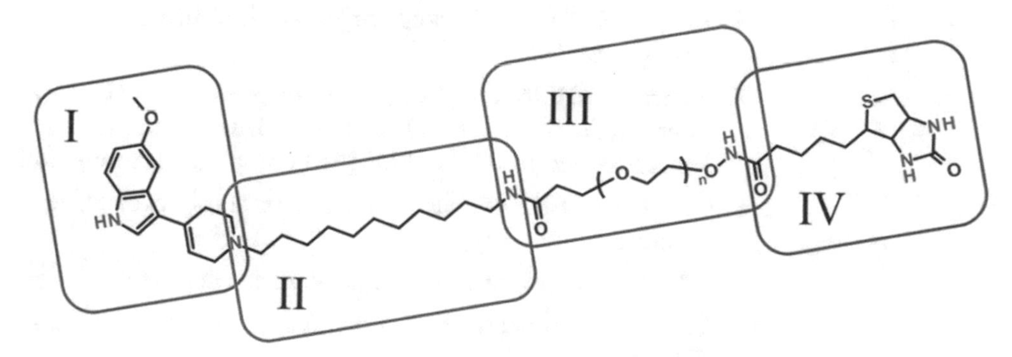

Fig. 1 Structure schematic of tailored organic ligands targeting plasma membrane monoamine transporters. Reprinted with permission from ref. 21. Copyright 2011 American Chemical Society

transporter structure and facilitates pseudo-irreversible binding, (2) a hydrophobic alkyl spacer which permits sufficient flexibility and provides a hydrophobic interface for a successful drug-binding pocket interaction, (3) a PEG polymer that aids in aqueous solubility and abolishes possible nonspecific interactions with the plasma membrane, and (4) a biotin moiety that allows subsequent streptavidin-conjugated QD recognition upon the transporter binding event (Fig. 1) [16–24].

In this chapter, we outline three fluorescence-based techniques that have been successfully applied to measure spatiotemporal changes in NTT localization and to establish dynamic imaging of individual NTT molecules using our ligand-conjugated QD approach. First, we discuss how to label and image membrane NTTs in live cells using QD probes in conjunction with ensemble fluorescence microscopy. Second, we present a more quantitative, flow cytometry-based protocol, particularly useful for assessing transporter internalization and recycling. Third, as dynamic trafficking of NTTs in the plasma membrane appears to be an important post-translational regulatory mechanism, we describe a single-molecule microscopy labeling protocol for determining the mobility of QD-bound transporters in the plasma membrane of live cells.

2 Materials

2.1 HEK293 Cell Culture and Reagents

1. DMEM medium (Gibco, Invitrogen Life Science, Bethesda, MD).
2. Phenol Red-free DMEM medium (Gibco, Invitrogen Life Science, Bethesda, MD).
3. Fetal bovine serum (FBS) (Gemini Bio-Products, West Sacramento, CA).
4. 0.05% Trypsin/EDTA (Cellgro, Mediatech).
5. L-Glutamine (Gibco, Invitrogen Life Science, Bethesda, MD).
6. T25/T75 flasks; 24-well or 96-well culture plates (BD Biosciences, Falcon).
7. Penicillin (10,000 U/mL) and streptomycin (10 mg/mL) solutions are frozen at −20 °C (Gibco, Invitrogen Life Science); 5 mL is added to 0.5 L of DMEM complete culture medium.
8. Cell line: HEK293 cells transiently or stably expressing NTT of interest.
9. 0.1 mg/mL poly-d-lysine solution in sterile H_2O.
10. Lab-Tek chambered #1.0 borosilicate coverglass (eight-well chamber).
11. Biotinylated ligand (1 mM stock solution in sterile H_2O stored dessicated at −20 °C).
12. Bovine serum albumin (BSA).
13. Streptavidin-conjugated quantum dots (SavQD605), with the emission maximum at 605 nm (Invitrogen Life Science, Bethesda, MD). Optimal filter is HQ 605/20 emission for QD605. QDs can be excited at any wavelengths, but 488 nm is a commonly utilized excitation line to minimize photodamage, QD blinking, and spectral cross-talk.
14. Cell Stripper, nonenzymatic dissociation buffer (Gibco, Invitrogen Life Science, Bethesda, MD).

15. Cell culture incubator, 37 °C, 5 % CO_2.
16. Vacuum pump for cell washes.

2.2 Equipment, Software, and Accessories

1. LSM 510 (Carl Zeiss) or LSM 710 (Carl Zeiss) equipped with a 63× 1.4 NA Apochromat oil-immersion objective lens and a 488-nm excitation line (Ar laser or solid-state diode laser); LSM 510/710 image acquisition/analysis software or ImageJ (NIH image analysis freeware) to process time-lapse and z-stack fluorescence images; microscope-mounted environmental chamber.
2. 3- or 5-laser Becton-Dickinson (BD Biosciences, San Jose, CA) bench-top flow cytometer equipped with a multiwall plate sample cube; 12×75 mm polystyrene flow cytometry tubes (BD Biosciences, San Jose, CA); FlowJo flow cytometry data analysis package (TreeStar, Ashland, OR).
3. High-speed, line-scanning Zeiss 5 Live confocal microscope equipped with a 63× 1.4 NA oil-immersion objective lens and a 488-nm 100-mW solid-state diode laser; microscope-mounted environmental chamber; Zeiss LSM Image Examiner software; MatLab or IDL-based programming routines for analysis of real-time QD trajectory data.
4. Imaging medium: phenol red-free DMEM supplemented with 1 % BSA.

3 Methods

For the purpose of this chapter, it is assumed that the NTT of interest is expressed in HEK293 cells; however, the general principles and protocols described below remain valid for any expression system being used.

3.1 Ensemble Microscopy Protocol

1. HEK293 cells are cultured in DMEM supplemented with 10 % FBS, 2 mM l-glutamine, 100 units/mL penicillin, and 100 mg/mL streptomycin and maintained at 37 °C with 5 % CO_2. For ensemble imaging, cells are plated in poly-d-lysine-treated (1 h at 37 °C) eight-well Lab-Tek chambered coverglass at a density of 1×10^5 to 1×10^6 cells/mL and cultured for 24 h prior to imaging.
2. Prior to labeling, wash the cells gently three times with warm phenol red-free culture medium.
3. Incubate cells with a biotinylated ligand (0.1–1 μM) in phenol red-free DMEM for 5–20 min at 37 °C. In the meantime, prepare a SavQD605 labeling by diluting SavQD605 stock solution in warm imaging buffer to reach a desired concentration of (0.5–2 nM) and incubate it in a 37 °C water bath for 10 min.
4. Wash the cells three times with warm phenol red-free DMEM.

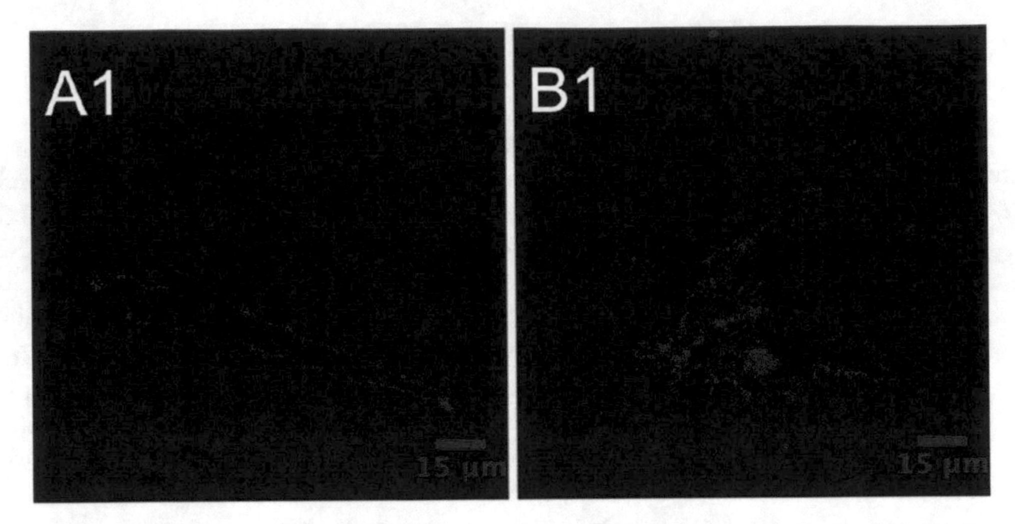

Fig. 2 Labeling of dopamine transporter (DAT) with ligand-conjugated QDots in live cells. (*left*) Streptavidin-conjugated QDots were used to label DATs previously exposed to a biotinylated, PEGylated cocaine analog. (**a1**) QD labeling of membrane DATs in a live HeLa cell. (**b1**) QD-bound DATs underwent acute redistribution from the plasma membrane to intracellular compartments as a result of protein kinase C (PKC) activation. Reprinted with permission from ref. 20. Copyright 2011 American Chemical Society

5. Incubate the cells with SavQD605 solution for 5 min at 37 °C and wash at least three times with warm imaging buffer.
6. Immediately post-labeling, place the chambered coverglass on the microscope with the mounted environmental chamber.
7. Acquire fluorescent images at 37 °C. Example data are shown in Fig. 2.

3.2 Flow Cytometry Protocol

1. Cells are plated in a poly-d-lysine-treated (1 h at 37 °C) 24-well/96-well culture plate at a density of $1–5 \times 10^5$ cells/mL 48 h prior to the flow cytometry assay.
2. Prior to QD conjugate labeling, wash the cells three times with warm DMEM and incubate with a drug for 10–30 min in complete culture medium at 37 °C and 5% CO_2. Parallel control wells are incubated with either drug-free complete culture medium (positive control) or in the presence of a high-affinity transporter inhibitor (negative control).
3. Wash the cells three times with warm phenol red-free DMEM and incubate with biotinylated ligand (0.1–1 μM)/drug mixture for 10 min at 37 °C in warm phenol red-free DMEM.
4. Wash the cells three times with ice-cold imaging buffer and incubate with previously prepared SavQD605 labeling solution in ice-cold imaging buffer.
5. Wash the cells gently three times with the ice-cold imaging buffer and add Cell Stripper solution. Incubate for 5–10 min at 37 °C.

6. Analyze cell QD fluorescence using a flow cytometer.
7. Data are typically collected from >10,000 cells per sample, with median fluorescence intensity as one of the recorded fluorescent signal parameters.
8. By utilizing median fluorescence intensity (MFI) parameter obtained from control cell populations, it is possible to compute the percentage of DAT molecules unavailable for binding (PI, percent inhibition) according to the equation below:

$$PI = \frac{MFI_{pos} - MFI_{treated}}{MFI_{pos} - MFI_{neg}} \times 100\,\%$$

where MFI_{pos} is MFI of a positive control (QD-ligand-labeled cells), MFI_{neg} is MFI of a negative control (QD only-labeled cells), and $MFI_{treated}$ is MFI of a cell population incubated with a certain DAT modulator dose and subsequently labeled with ligand-conjugated QDs [23]. Example data are shown in Fig. 3.

3.3 Single-Molecule Microscopy Protocol

1. Cells are plated in poly-d-lysine-treated (1 h at 37 °C) eight-well Lab-Tek chambered coverglass at a density of $1–5 \times 10^4$ cells/mL and cultured for 24 h prior to imaging.
2. Prior to labeling, wash the cells gently three times with warm phenol red-free culture medium.
3. Incubate cells with a biotinylated ligand (1–100 nM) in phenol red-free DMEM for 5–20 min at 37 °C. In the meantime, prepare a SavQD605 labeling by diluting SavQD605 stock solution in warm imaging buffer to reach a desired concentration of (0.01–0.5 nM) and incubate it in a 37 °C water bath for 10 min.
4. Wash the cells three times with warm phenol red-free DMEM.
5. Incubate the cells with SavQD605 solution for 5 min at 37 °C and wash thoroughly at least three times with warm imaging buffer to remove unbound QDs.
6. Immediately post-labeling, place the chambered coverglass on the microscope with the mounted environmental chamber.
7. Acquire time-lapse fluorescent images at 37 °C immediately after the final wash step. Typically, the final wash step is carried in the immediate vicinity of the imaging system.
8. Live imaging should not exceed 30 min at 37 °C for cell survival and is optimally carried out within the initial 10–15 min to limit turnover of QD-bound membrane NTT molecules.
9. Real-time, time-lapse image recording is obtained with an integration time of 25–100 ms for at least 500 consecutive frames. Example series are shown in Fig. 4.

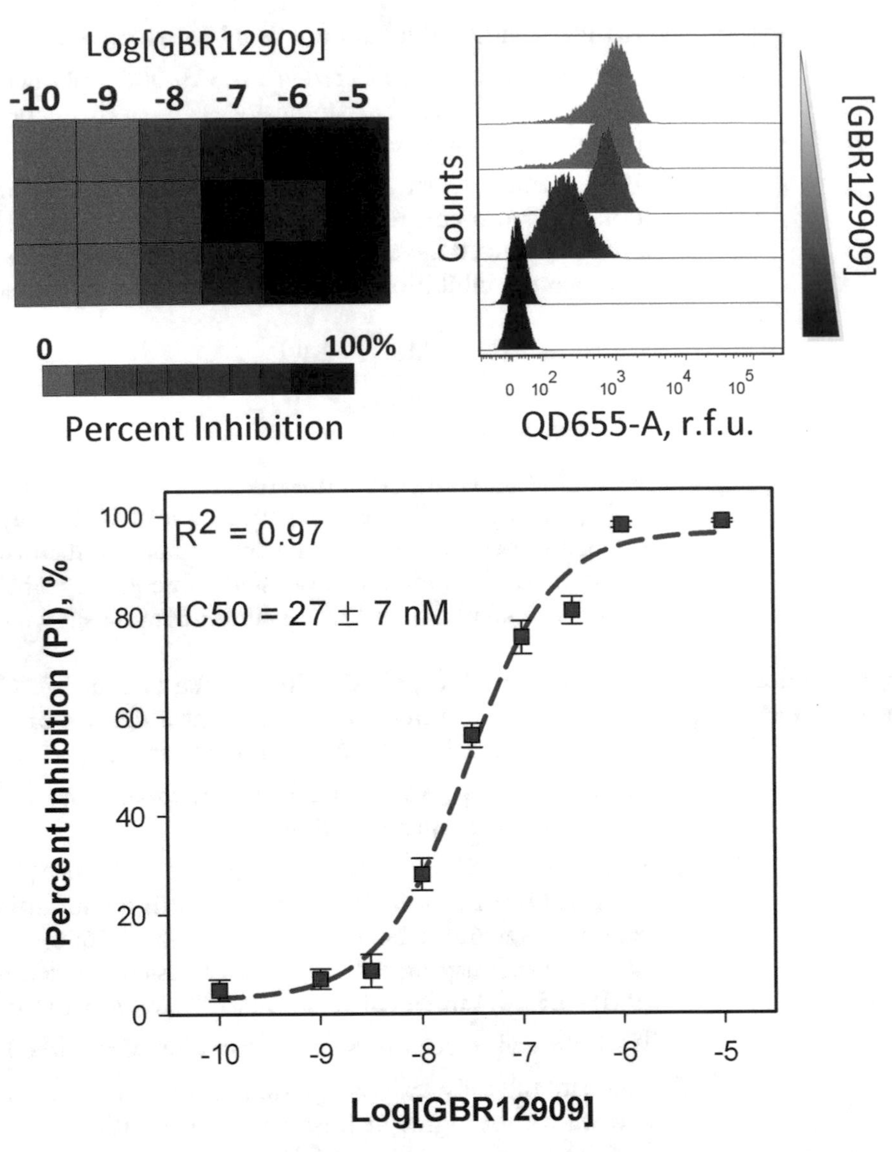

Fig. 3 Flow cytometry-based screening of the inhibitory activity of GBR12909, a high-affinity DAT antagonist, using antagonist-conjugated Qdots. DAT-expressing HEK cells were treated with five- or tenfold dilutions of GBR12909. Percent inhibition at increasing doses of GBR12909 is represented as a heat map (*top left*) and representative histogram plots of the effects of increasing GBR12909 doses (*top right*) on QD conjugate binding are shown. The heat map and IC50 curve (*bottom*) were generated using median QD fluorescence intensity values. Reprinted with permission from ref. 23. Copyright 2012 Royal Chemical Society

10. Real-time trajectory data is subsequently obtained from the recorded time-lapse image series and analyzed using custom programs written in Matlab or IDL programming software. Tracking analysis sequence is illustrated in Fig. 5.

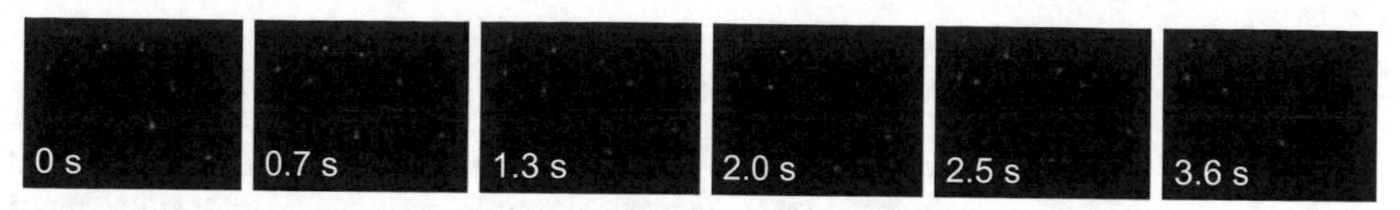

Fig. 4 Time-lapse image series depicting movement of cell surface QD-bound transporters

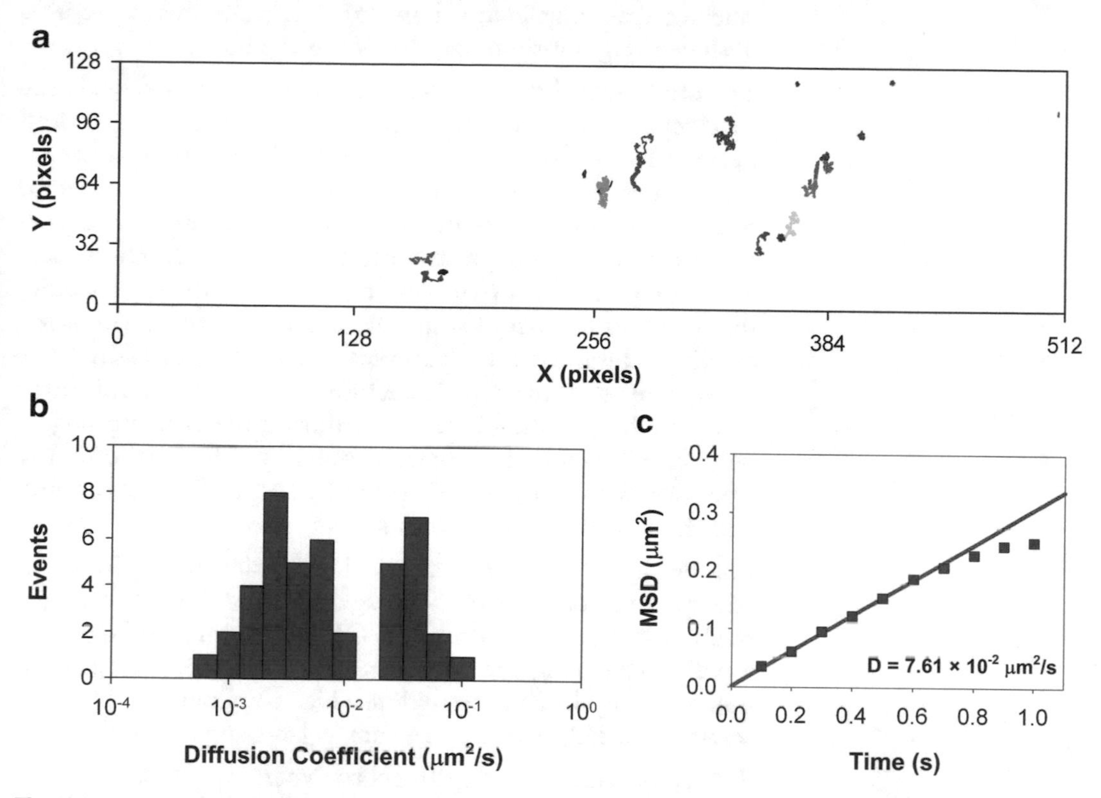

Fig. 5 Schematic illustrating trajectory data analysis in a typical single-QD tracking experiment. (**a**) Example of QD-DAT trajectories on the surface of transfected HEK293 cells. Scale: 1 pixel = 200 nm. (**b**) A histogram showing diffusion coefficients determined for the trajectories in **a**. (**c**) Averaged mean square displacement (MSD)-time plot of QD trajectories. Ensemble diffusion coefficient is estimated via the linear fit of 2–5 MSD-time plot data points

4 Notes

1. Optimal plating density and cell health are critical to keep weakly adherent cells, such as HEK293, from detaching off the Lab-Tek chambered coverglass throughout the protocol. Treatment of the chambered coverglass with poly-d-lysine solution is a necessary step to ensure the cells remain adhered to the glass bottom through the extensive series of incubation and wash steps. Also, it is of utmost importance that one carefully examines cell morphology and overall cell health prior to acquiring fluorescence data.

2. One of the most important variables for a successful experiment is adequate quality and quantity of washings after drug, biotinylated ligand, and QD incubation. One must wash extensively after each separate step to remove excess, unbound probes, as they have the potential to interfere with subsequent recognition events and ultimately affect the specificity of QD labeling. Additionally, it is imperative that one always prepares fresh working solutions the day of the experiment.
3. The most critical determinant of experimental success is the specificity of biotinylated ligand binding. One must find optimal ligand concentration and incubation time to maximize specific binding. In our experience, 0.1–1 μM and 5–20 min ranges for ligand dose and incubation time respectively are typically a good starting point. In all cases, one must run parallel control samples to ensure labeling specificity. The control samples usually are to (1) apply the same labeling conditions to parental cells not expressing the transporter of interest, (2) include a high-affinity inhibitor to block the specific binding site during the labeling protocol, and (3) label transporter-expressing cells with only the QD probes to assess the degree of nonspecific QD binding and the effectiveness of wash steps.
4. Another important aspect of assuring labeling specificity is the addition of a common blocking agent, such as BSA, to the QD solution and the imaging buffer. QD nonspecific binding varies significantly among cell types, and one must take great care to optimize the blocking conditions (Fig. 6). Common blocking reagents are BSA, FBS, horse serum, gelatin, and nonfat dry milk.
5. As QD-bound NTTs are subject to dynamic protein turnover, fluorescence data acquisition must be conducted immediately after the final wash step, especially in the case of single-

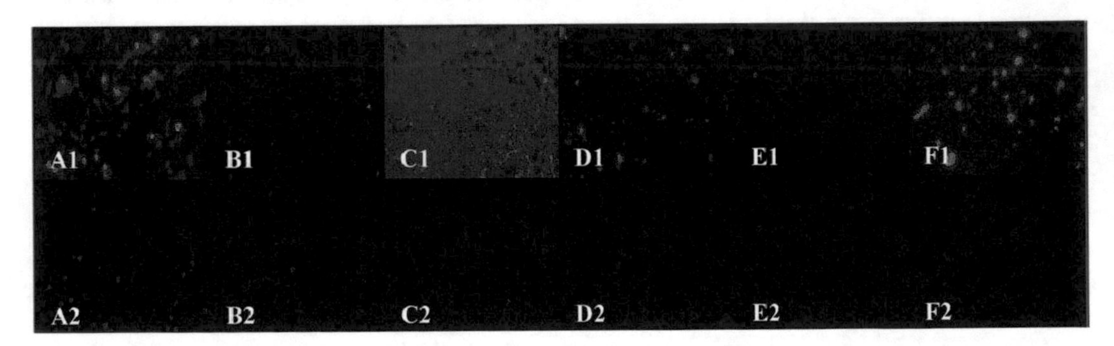

Fig. 6 Comparison of nonspecific cell surface binding of 50 nM AMP™ Dots (**a1–f1**) and PEGylated AMP™ Dots (**a2–f2**). Nonspecific binding was found to be dependent upon the cell type, and conjugation of methoxy-terminated PEG2000 to the surface of AMP™ Dots resulted in significant reduction of nonspecific cellular binding. Figure reproduced with permission from ref. 40. Copyright 2005 American Chemical Society

molecule experiments. This helps prevent transporter endocytosis and allows adequate visualization of membrane-restricted signaling events.

6. An important consideration is controlling the valency of the binding. This is a particularly critical parameter in single-molecule experiments, as multivalent QD labeling leads to protein cross-linking that may inadvertently trigger downstream signal transduction cascades. To this end, there are two common solutions. One involves preincubation of SavQDs with the biotinylated ligand at ~1:1 ratio in a borate buffer (pH ~8.5) for 0.5–24 h at room temperature with constant stirring; the other involves a two-step labeling protocol as described above and the use of ~equimolar doses of biotinylated ligand and SavQDs. In the case of endogenous expression systems, this requirement can be relaxed, as the low surface density of transporters is the primary determinant of monovalent labeling.
7. QD density must be adjusted accordingly to ensure maximum signal-to-noise ratio in ensemble imaging and permit observation of 10–20 individual QDs on the cell surface in a single-molecule experiment. This is achieved via titrating the QD concentration while keeping the concentration of biotinylated ligand constant.
8. Table 3 provides a set of troubleshooting instructions for a typical single-molecule experiment [39].

Table 3

Troubleshooting a single-QD imaging experiment

Problem	Cause	Solution
Low or excessive QD label density	Inappropriate biotinylated ligand concentration	Optimize ligand concentrations for labeling
	Poor or excessive transporter expression	Check whether the transporter is delivered to the surface. Optimize protein expression level
	QD aggregation	Prepare fresh QD dilution and store for no longer than a few hours
	Poor ligand affinity	Check ligand affinity via transport uptake assay
Nonspecific labeling	Excessive ligand or QD concentration	Reduce and optimize ligand and QD concentrations
Excessive QD blinking	Excessive excitation duration or intensity	Minimize excitation intensity, use longer wavelengths, and reduce the excitation duration

Acknowledgements

The authors wish to thank Prof. Randy D. Blakely, Dr. Jerry C. Chang, and Dr. Ian D. Tomlinson for all the helpful discussions and suggestions. This work was supported by grants from National Institutes of Health EB003728 to S.J.R. O.K. would also like to acknowledge Vanderbilt Institute for Nanoscale Science and Engineering (VINSE) fellowship.

References

1. Torres GE, Gainetdinov RR, Caron MG (2003) Plasma membrane monoamine transporters: structure, regulation and function. Nat Rev Neurosci 4(1):13–25
2. González MI, Robinson MB (2004) Neurotransmitter transporters: why dance with so many partners? Curr Opin Pharmacol 4(1):30–35
3. Sager JJ, Torres GE (2011) Proteins interacting with monoamine transporters: current state and future challenges. Biochemistry 50(34): 7295–7310
4. Fei H, Grygoruk A, Brooks ES, Chen A, Krantz DE (2008) Trafficking of vesicular neurotransmitter transporters. Traffic 9(9):1425–1436
5. Ramamoorthy S, Shippenberg TS, Jayanthi LD (2011) Regulation of monoamine transporters: role of transporter phosphorylation. Pharmacol Ther 129(2):220–238
6. Haraguchi T (2002) Live cell imaging: approaches for studying protein dynamics in living cells. Cell Struct Funct 27(5): 333–334
7. Sako Y, Yanagida T (2003) Single-molecule visualization in cell biology. Nat Rev Mol Cell Biol (Suppl): SS1–SS5
8. Resch-Genger U, Grabolle M, Cavaliere-Jaricot S, Nitschke R, Nann T (2008) Quantum dots versus organic dyes as fluorescent labels. Nat Methods 5(9):763–775
9. Lippincott-Schwartz J, Patterson GH (2003) Development and use of fluorescent protein markers in living cells. Science 300(5616): 87–91
10. Zhang J, Campbell RE, Ting AY, Tsien RY (2002) Creating new fluorescent probes for cell biology. Nat Rev Mol Cell Biol 3(12):906–918
11. Alivisatos AP, Gu W, Larabell C (2005) Quantum dots as cellular probes. Annu Rev Biomed Eng 7:55–76
12. Bruchez M Jr, Moronne M, Gin P, Weiss S, Alivisatos AP (1998) Semiconductor nanocrystals as fluorescent biological labels. Science 281(5385):2013–2016
13. Chan WC, Nie S (1998) Quantum dot bioconjugates for ultrasensitive nonisotopic detection. Science 281(5385):2016–2018
14. Rosenthal SJ, Chang JC, Kovtun O, McBride JR, Tomlinson ID (2011) Biocompatible quantum dots for biological applications. Chem Biol 18(1):10–24
15. Chang JC, Kovtun O, Blakely RD, Rosenthal SJ (2012) Labeling of neuronal receptors and transporters with quantum dots. Wiley Interdiscip Rev Nanomed Nanobiotechnol 4(6):605–619
16. Rosenthal SJ, Tomlinson I, Adkins EM, Schroeter S, Adams S, Swafford L, McBride J, Wang Y, DeFelice LJ, Blakely RD (2002) Targeting cell surface receptors with ligand-conjugated nanocrystals. J Am Chem Soc 124(17):4586–4594
17. Tomlinson ID, Mason JN, Blakely RD, Rosenthal SJ (2005) Inhibitors of the serotonin transporter protein (SERT): the design and synthesis of biotinylated derivatives of 3-(1,2,3,6-tetrahydro-pyridin-4-yl)-1H-indoles. High-affinity serotonergic ligands for conjugation with quantum dots. Bioorg Med Chem Lett 15(23):5307–5310
18. Tomlinson ID, Mason JN, Blakely RD, Rosenthal SJ (2006) High affinity inhibitors of the dopamine transporter (DAT): novel biotinylated ligands for conjugation to quantum dots. Bioorg Med Chem Lett 16(17):4664–4667
19. Tomlinson ID, Chang J, Iwamoto H, Felice LJD, Blakely RD, Rosenthal SJ (2008) Targeting the human serotonin transporter (hSERT) with quantum dots. SPIE 6866:68660X
20. Kovtun O, Tomlinson ID, Sakrikar DS, Chang JC, Blakely RD, Rosenthal SJ (2011) Visualization of the cocaine-sensitive dopamine transporter with ligand-conjugated quantum dots. ACS Chem Neurosci 2:370–378
21. Chang JC, Tomlinson ID, Warnement MR, Iwamoto H, DeFelice LJ, Blakely RD, Rosenthal SJ (2011) A fluorescence displacement assay for antidepressant drug discovery based on ligand-conjugated quantum dots. J Am Chem Soc 133(44):17528–17531

22. Tomlinson ID, Iwamoto H, Blakely RD, Rosenthal SJ (2011) Biotin tethered homotryptamine derivatives: high affinity probes of the human serotonin transporter (hSERT). Bioorg Med Chem Lett 21(6): 1678–1682
23. Kovtun O, Ross EJ, Tomlinson ID, Rosenthal SJ (2012) A flow cytometry-based dopamine transporter binding assay using antagonist-conjugated quantum dots. Chem Commun 48(44):5428–5430
24. Chang JC, Tomlinson ID, Warnement MR, Ustione A, Carneiro AMD, Piston DW, Blakely RD, Rosenthal SJ (2012) Single molecule analysis of serotonin transporter regulation using antagonist-conjugated quantum dots reveals restricted, p38 MAPK-dependent mobilization underlying uptake activation. J Neurosci 32(26):8919–8929
25. Fjorback AW, Pla P, Müller HK, Wiborg O, Saudou F, Nyengaard JR (2009) Serotonin transporter oligomerization documented in RN46A cells and neurons by sensitized acceptor emission FRET and fluorescence lifetime imaging microscopy. Biochem Biophys Res Commun 380(4):724–728
26. Schmid JA, Scholze P, Kudlacek O, Freissmuth M, Singer EA, Sitte HH (2001) Oligomerization of the human serotonin transporter and of the rat GABA transporter 1 visualized by fluorescence resonance energy transfer microscopy in living cells. J Biol Chem 276(6):3805–3810
27. Furman CA, Chen R, Guptaroy B, Zhang M, Holz RW, Gnegy M (2009) Dopamine and amphetamine rapidly increase dopamine transporter trafficking to the surface: live-cell imaging using total internal reflection fluorescence microscopy. J Neurosci 29(10):3328–3336
28. Egaña LA, Cuevas RA, Baust TB, Parra LA, Leak RK, Hochendoner S, Peña K, Quiroz M, Hong WC, Dorostkar MM, Janz R, Sitte HH, Torres GE (2009) Physical and functional interaction between the dopamine transporter and the synaptic vesicle protein synaptogyrin-3. J Neurosci 29(14):4592–4604
29. Grånäs C, Ferrer J, Loland CJ, Javitch JA, Gether U (2003) N-terminal truncation of the dopamine transporter abolishes phorbol ester- and substance P receptor-stimulated phosphorylation without impairing transporter internalization. J Biol Chem 278(7):4990–5000
30. Sorkina T, Richards TL, Rao A, Zahniser NR, Sorkin A (2009) Negative regulation of dopamine transporter endocytosis by membrane-proximal N-terminal residues. J Neurosci 29(5):1361–1374
31. Rao A, Richards TL, Simmons D, Zahniser NR, Sorkin A (2012) Epitope-tagged dopamine transporter knock-in mice reveal rapid endocytic trafficking and filopodia targeting of the transporter in dopaminergic axons. FASEB J 26(5):1921–1933
32. Rao A, Simmons D, Sorkin A (2011) Differential subcellular distribution of endosomal compartments and the dopamine transporter in dopaminergic neurons. Mol Cell Neurosci 46(1):148–158
33. Hadrich D, Berthold F, Steckhan E, Bönisch H (1999) Synthesis and characterization of fluorescent ligands for the norepinephrine transporter: potential neuroblastoma imaging agents. J Med Chem 42(16):3101–3108
34. Cha JH, Zou M-F, Adkins EM, Rasmussen SGF, Loland CJ, Schoenenberger B, Gether U, Newman AH (2005) Rhodamine-labeled 2β-carbomethoxy-3β-(3,4-dichlorophenyl)tropane analogues as high-affinity fluorescent probes for the dopamine transporter. J Med Chem 48(24):7513–7516
35. Eriksen J, Rasmussen SGF, Rasmussen TN, Vaegter CB, Cha JH, Zou M-F, Newman AH, Gether U (2009) Visualization of dopamine transporter trafficking in live neurons by use of fluorescent cocaine analogs. J Neurosci 29(21):6794–6808
36. Zhang P, Jørgensen TN, Loland CJ, Newman AH (2013) A rhodamine-labeled citalopram analogue as a high-affinity fluorescent probe for the serotonin transporter. Bioorg Med Chem Lett 23(1):323–326
37. Li M, Lester HA (2002) Early fluorescence signals detect transitions at mammalian serotonin transporters. Biophys J 83(1):206–218
38. Zhao Y, Terry D, Shi L, Weinstein H, Blanchard SC, Javitch JA (2010) Single-molecule dynamics of gating in a neurotransmitter transporter homologue. Nature 465(7295):188–193
39. Bannai H, Levi S, Schweizer C, Dahan M, Triller A (2007) Imaging the lateral diffusion of membrane molecules with quantum dots. Nat Protoc 1(6):2628–2634
40. Bentzen EL, Tomlinson ID, Mason J, Gresch P, Warnement MR, Wright D, Sanders-Bush E, Blakely R, Rosenthal SJ (2005) Surface modification to reduce nonspecific binding of quantum dots in live cell assays. Bioconjug Chem 16(6):1488–1494

Chapter 9

Approaches for Investigating Transporter Endocytic Trafficking in Cell Lines and Native Preparations

Haley E. Melikian, Sijia Wu, and Luke R. Gabriel

Abstract

Neurotransmitter transporters are key determinants of synaptic strength and studies over the past two decades have clearly demonstrated that compromised transporter function, either via genetic manipulation or intrinsic to neuropsychiatric disorders, has a profound impact on behavior and synaptic homeostasis. Transporters are dynamically regulated by numerous intracellular signaling pathways, many of which acutely alter transporter plasma membrane expression via endocytic trafficking. Given the importance of transport function in synaptic function, copious attention has been devoted to understanding the molecular mechanisms governing transporter cell surface stability. In this chapter, we provide a comprehensive overview of the available methods used to probe constitutive and regulated transporter endocytic trafficking both in cultured cells and ex vivo brain slices. Detailed protocols are provided and both the advantages and limitations of the various methodologies are considered.

Key words Transport, Reuptake, Trafficking, Endocytosis, Regulation

1 Introduction

For the past 50 years, neurotransmitter transport function in synaptic terminals has been widely studied in a variety of contexts, both as the principal targets of therapeutic and addictive drugs, as well as responders to environmental stressors [1–3]. Membrane binding and synaptosomal preparations confirmed that neurotransmitter transporters are expressed at the plasma membrane and ultrastructural analyses further revealed transporter localization to perisynaptic regions [4–6]. Surprisingly, these studies also discovered robust transporter pools in intracellular membranous compartments whose function was initially unclear. Over the past decade, it has become increasingly evident that neurotransmitter transporters are not static in the plasma membrane. Rather, they are dynamically trafficked to and from the cell surface via endocytic trafficking, and substantial transporter endosomal pools rapidly regulate transporter cell surface presentation [2, 3, 7]. Transporter

Heinz Bönisch and Harald H. Sitte (eds.), *Neurotransmitter Transporters: Investigative Methods*, Neuromethods, vol. 118, DOI 10.1007/978-1-4939-3765-3_9, © Springer Science+Business Media New York 2016

trafficking is integral to transporter regulatory events, both in response to cellular signaling cascades and exposure to psychostimulants [2, 7–12]. In addition, recent studies have identified transporter coding variants in patient pedigrees that exhibit aberrant trafficking, and which underlie transporter dysregulation [13, 14].

The growing interest in transporter membrane trafficking has driven the development of tools to examine basal and regulated transporter trafficking both in cultured cells and in ex vivo preparations. Here, we present a variety of biochemical and immunocytochemical approaches for investigating transporter trafficking, and discuss the advantages and pitfalls of each.

2 Biochemical Approaches

The most widely used method for examining surface and trafficking transporter pools is to covalently couple biotin to extracellularly exposed primary amines at the cell surface using sulfo-*N*-hydroxysuccinyl (NHS) esters. These reagents offer a number of distinct advantages. They are membrane impermeant, commercially available, affordable and easy to use. They are also available in a variety of forms, including those with linkers between the reactive ester and biotin, such as cleavable disulfides and spacers regions, thus affording variations in approaches to suit different experimental conditions. After labeling, biotinylated (cell surface) proteins are separated from non-biotinylated (intracellular) proteins by batch streptavidin chromatography and transporter distributions in these populations are identified by immunoblotting for proteins of interest. Transporter surface levels are normalized to the total amount of cellular transporter, determined either from total cell lysates or from the residual non-biotinylated fraction. Another distinct advantage to this approach is the high degree of versatility, wherein slight protocol variations can extend the utility beyond simple steady state measurements, to allow for measuring trafficking kinetics, such as internalization and recycling rates. Moreover, since the majority of surface proteins are biotinylated during the experiment, specificity controls are conveniently "built-in" to each experiment.

2.1 Steady State Biotinylation in Adherent Cultured Cells

This is a widely used method to measure changes in wild type or mutant transporter surface levels in response to a stimulus. Cells are treated with a given stimulus under trafficking permissive temperatures (37 °C), rapidly chilled to stop all endocytic trafficking and the cell surface protein population is biotinylated. Thus, relative changes in surface levels can be compared between control and experimental conditions. This method is highly accurate in homogenous cell populations, but is less reliable in mixed cell populations (e.g., Cells transfected with more than one protein or during protein knockdown, where varying expression levels of the proteins of interest may skew the results across the population).

3 Required Reagents

1. *PBS^{2+}*

   Phosphate-buffered saline, pH 7.4, supplemented with 1.0 mM $MgCl_2$, 0.1 mM $CaCl_2$.
2. *Quench Solution*

   PBS^{2+} supplemented with 100 mM glycine.
3. *RIPA Lysis Buffer*

   10 mM Tris, pH 7.4, 150 mM NaCl, 1.0 mM EDTA, 1% Triton X-100, 0.1% SDS, 1% Na deoxycholate.
4. *RIPA with Protease Inhibitors* (*RIPA/PI*)—Make fresh daily

   RIPA supplemented with 1 μM leupeptin, 1 μM pepstatin, 1 μM aprotinin, and 1 mM phenylmethyl sulfonyl fluoride.
5. *"Happy" PBS*—Make fresh daily

   PBS^{2+}, 0.18% (w/v) glucose, 0.2% IgG-free, protease-free BSA (Sigma-Aldrich #A3059).
6. *Sulfo-N-hydroxysuccinyl-SS-Biotin* (*sulfo-NHS-SS-biotin, Pierce Chemical Company*)

   Stock solutions should be 200 mg/ml in DMSO and are resistant to multiple freeze–thaw cycles. Aliquots are stored at −20 °C. The succinyl ester is rapidly hydrolyzed in aqueous solution, so working solutions should be prepared immediately prior to applying to cells.
7. *Streptavidin Agarose* (*Pierce Chemical Company*)

   Provided as a 50% slurry. Mix well prior to sampling.

4 Protocol

4.1 Cell Plating

1. *Cell lines—no transfection*: Plate cells to 70–80% density in 6-well culture dishes 1 day prior to assaying. One well of a 6-well plate per sample is usually sufficient, at a cell density of 1×10^6 cells/well. If cells are not strongly adherent, use plates coated with an appropriate cell adhesion substrate, such as poly-d-lysine.
2. *Cell lines—with transfection*: Plate cells and transfect according to the optimized protocol, such that they will be ~70–80% confluent on the assay day.
3. *Primary cultured cells*: Use sufficient numbers to assure that transporter of interest will be detectable by immunoblot.

4.2 Drug Treatment (If Appropriate) and Biotinylation

Make sure all solutions are ice cold. Perform all biotinylation steps in cold room, using ice-cold reagents and incubating plates on an ice bath.

1. Wash cells rapidly three times in prewarmed "Happy" PBS.
2. Treat cells with the experimental stimuli of interest.
3. Shift cells to trafficking restrictive temperature.
   (a) Place on an ice bath in cold room.
   (b) Wash cells 4 × 2 ml w/ice-cold PBS^{2+}.
   (c) Drain all residual buffer by tilting plates and aspirating.
4. Make 1 mg/ml sulfo-NHS-SS-biotin in PBS^{2+}.
   (a) Prepare a sufficient amount to evenly coat the bottom of each well.
   (b) 750 μl/well is sufficient for one well of a 6-well culture plate.
5. Add 750 μl sulfo-NHS-SS-biotin per well and incubate for 15 min, at 4 °C on ice, shaking vigorously to assure complete coverage of the well.
6. Make another batch of 1 mg/ml sulfo-NHS-SS-biotin in PBS^{2+}, aspirate and discard old solution and replace wells w/ fresh solution. Incubate for 15 min, at 4 °C on ice, shaking.
7. Wash rapidly three times with 2.0 ml/well ice-cold quench solution.
8. Quench residual reactive NHS groups by incubating twice in 1–2 ml quench solution/well × 15 min, 4 °C on ice, shaking.

4.3 Prepare Cell Lysate

1. Aspirate quench solution.
2. Wash cells three times with 2.0 ml PBS^{2+}/well to remove residual glycine.
3. Lyse cells in 250–300 μl ICE-COLD RIPA/PI, 20 min, 4 °C, shaking.
4. Remove cellular debris by centrifugation.
5. Transfer lysates to microcentrifuge tubes.
6. Centrifuge at 18,000 × *g*, for 10 min, at 4 °C.
7. Keep samples on ice until ready to use.

4.4 Determine Lysate Protein Concentration

1. BCA protein assay (Pierce, or other detergent-compatible assay system).
2. Use BSA standards in RIPA.
3. Be careful not to disturb insoluble pellets when sampling the lysates.

4.5 Isolate Biotinylated Proteins

1. Optimize bead–lysate ratio.

Note: Transporter expression levels can vary widely in different cell types. Therefore, prior to initiating a biotinylation study in a new model system, it is essential to empirically determine the optimal bead–lysate ratio for a given cell type, in order to quantitatively recover all biotinylated proteins from a given amount of cell lysate. If this step is omitted, it is impossible to accurately determine putative surface level changes, due to incomplete isolation of biotinylated proteins.

1. Incubate 25 μl streptavidin agarose beads with increasing amounts of cell lysate (approximate range 25–200 μg).
2. Proceed as described below for binding, washing and elution steps.
3. Quantify the resulting Immunoreactive bands.
4. Choose a bead–lysate ratio in the linear range of binding that permits accurate quantification of either increased or decreased transporter surface expression.

4.6 Prepare Streptavidin Agarose Beads

1. Determine amount of beads you will need for all your samples. Prepare a sufficient bead volume for that amount plus one extra (i.e., four samples at 25 μl/sample = 100 μl beads + one extra = 125 μl beads).
2. Vortex streptavidin bead stock and pipette out desired volume. If using a P200 pipette, cut off the end of the tip to prevent blockage or bead damage
3. Wash beads three times in 0.5–1.0 ml RIPA/PI to remove preservatives. Vortex after adding RIPA and collect beads between washes by centrifuging at 18,000 × *g*, for 1 min, at room temperature. Aspirate off buffer between washes using a glass Pasteur pipette attached to a vacuum flask. We find that attaching a plastic P200 tip to the end of the glass Pasteur pipette provides finer control when aspirating. It is not necessary to remove absolutely all the buffer for the first two washes, as it risks aspirating beads into the pipette. After the final wash, remove as much excess buffer as possible without aspirating the beads.
4. Add RIPA/PI to bring beads back to their original volume, pipetting up and down to resuspend. Avoid vortexing at this point, as beads will stick to the tube wall.
5. Aliquot beads into microcentrifuge tubes using a P200 with the tip cut off. Pipette up and down several times between aliquots to assure beads remain evenly suspended and dispersed among the tubes.

4.7 Bind Biotinylated Proteins to Beads

1. Distribute cell lysates to the tubes containing the agarose beads. For experiments where multiple samples are being compared, make sure to use the same amount of protein for each sample for accurate comparisons.
2. Add additional RIPA/PI to bring to samples to a 200 μl minimal volume. This assures that samples will mix adequately during the incubation and also unifies the protein concentrations across samples.
3. Place tubes in a tube rotator and mix overnight at 4 °C.

4.8 Prepare Representative Total Samples

1. In separate tubes, dispense an equivalent of the total lysate volume used for each sample. Alternatively, if the sample volumes are high, a fraction of total lysate volume can be reserved instead, to accommodate maximal load volumes on SDS-PAGE gels. These will be used to normalize the surface expression to the total transporter amount.
2. Add either an equal volume of 2× or 1/5 volume of 6× SDS-PAGE sample buffer.
3. Store at −20 °C until samples are analyzed.

Note: In some cases, there is insufficient lysate available for the normalizing control. An alternative is to save the supernatant (intracellular protein) after isolating biotinylated proteins. Supernatants are spun concentrated in microcentrifuge (4 °C) and analyzed in parallel with biotinylated fractions. Total protein levels are subsequently determined by adding surface and intracellular fractions.

4.9 Wash Beads and Elute Samples

1. Centrifuge bead samples at 18,000 × *g*, for 2 min, at room temperature
2. Aspirate supernatant and wash beads three times with 0.75 ml RIPA. To minimize bead loss, leave a small head volume of buffer above beads between washes. After the final wash, remove as much RIPA as possible, without disrupting the bead pellet.
3. Elute biotinylated proteins from streptavidin beads by reducing the disulfide linkage. Add 25 μl 2× SDS-PAGE sample buffer, 100 mM DTT, vortex well, and rotate samples for 30 min, at room temperature.

Note: DO NOT BOIL TRANSPORTER SAMPLES PRIOR TO SDS-PAGE. Transporters have a high tendency to aggregate when boiled in SDS-PAGE sample buffer, resulting in inability to migrate into the SDS-PAGE gel. If boiling is absolutely necessary (i.e., if another protein to be monitored in parallel must be boiled), supplement sample buffer with 2 M urea (final concentration) to minimize aggregation. While effective in reducing aggregation, urea does compromise band appearances.

4.10 Analyze Samples

1. Thaw total lysate samples and rotate in parallel with bead samples, for 30 min, at room temperature.
2. Separate proteins by SDS-PAGE.
3. Immunoblot gels for transporter of interest. Probe blots in parallel for any housekeeping gene to assure even loading. Capture bands in linear range of detection for accurate quantification.
4. Quantify band densities and calculate relative transporter surface density as a % of the total transporter signal.

5 Representative Results

As widely reported, the dopamine transporter (DAT) is rapidly lost from the plasma membrane in response to amphetamine (AMPH) exposure [15–17] and PKC activation [18–22]. Moreover, work from our lab [19] and others [23] reveals that a negative endocytic regulatory mechanisms, or "endocytic brake", stabilizes DAT in the plasma membrane and is disengaged following PKC activation. We recently reported that the non-receptor tyrosine kinase, cdc42 effector, Ack1, is the penultimate mechanism that engages the DAT endocytic brake and stabilizes DAT at the plasma membrane [24]. We used highly specific, small molecule cdc42 inhibitors to ask (1) whether cdc42 activation stabilizes DAT in the plasma membrane and (2) whether cdc42 inactivation acts downstream of PKC to disengage the DAT endocytic brake. As seen in Fig. 1, SK-N-MC cells stably expressing DAT were pretreated ±cdc42 inhibitors casin (Fig. 1a) or pirl1 (Fig. 1b), 30 min, 37 °C, followed by treatment with either vehicle or 1 μM PMA, 30 min 37 °C, to activate PKC. DAT surface levels were subsequently measured by steady state biotinylation. PMA treatment significantly decreased surface DAT, as compared to vehicle-treated cells. Interestingly, treatment with either casin or pirl1 significantly decreased DAT surface expression and PKC activation had no further effect on DAT surface levels following cdc42 inactivation. These results exemplify how steady state biotinylation is capable of accurately and precisely measuring moderate, but significant, changes in transporter surface levels.

5.1 Steady State Biotinylation in Acute Brain Slices

This method boasts several advantages. Measurements are made ex vivo in intact adult neurons in contrast to cultured primary neurons, which are usually prepared from developmentally immature cells whose phenotype(s) may not accurately reflect that of fully developed adult neurons. Slices can be treated directly with drugs following harvesting, or can be harvested from whole animals following a variety of experimental conditions, including optogenetic modulation of neuronal activity, gene manipulation, establishing behaviors and/or whole animal or locally delivered

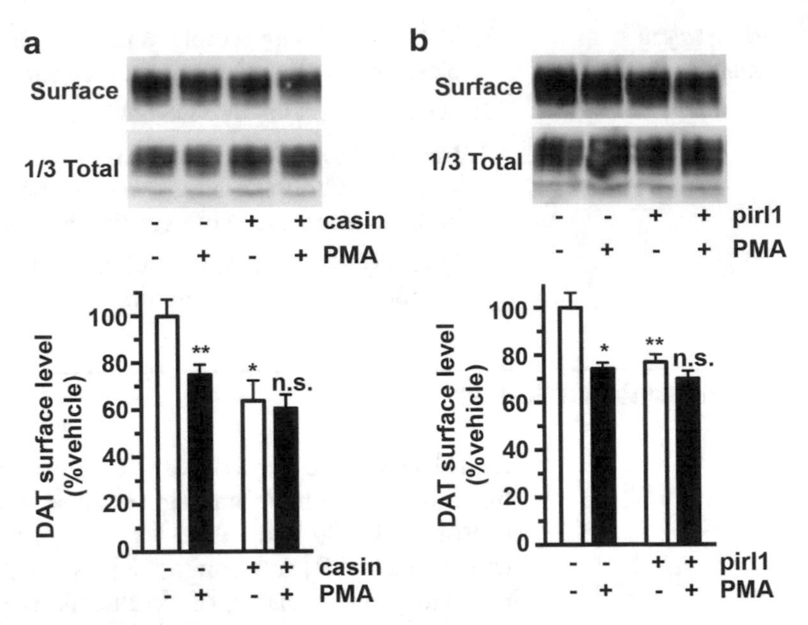

Fig. 1 Cdc42 stabilizes DAT surface expression. *Cell surface biotinylation:* DAT SK-N-MC cells were pretreated ±10 μM casin (**a**) or ±20 μM pirl1 (**b**), 30 min, 37 °C, followed by treatment ±1 μM PMA, 30 min, 37 °C. Relative DAT surface levels were measured by biotinylation as described in *Protocol*. Representative immunoblots are shown in the top of each panel. (**a**) Casin treatment. Average DAT surface levels expressed as %vehicle levels ± S.E.M. *Asterisks* indicate a significant difference from vehicle control, $*p < 0.05$, $**p < 0.01$, one-way ANOVA with Bonferroni's multiple comparison test, $n = 5–6$. (**b**) Pirl1 treatment. Average DAT surface levels expressed as %vehicle levels ± S.E.M. *Asterisks* indicate a significant difference from vehicle control, $*p < 0.02$, $**p < 0.01$, one-way ANOVA with Bonferroni's multiple comparison test, $n = 3$. *Adapted from* [24]

drug treatments. Additionally, region-specific differences in trafficking can be examined with this method. The major limitation of this approach is brain tissue viability, which precludes performing experiments that require either extended incubation periods or multiple rounds of tissue chilling and re-warming, which are not well tolerated by acute brain slices and which, in our hands, perturb normal trafficking. Thus, this approach is currently limited to acute trafficking effects that are detectable on a 10–60 min timescale, post slice preparation.

6 Required Reagents

6.1 Artificial Cerebrospinal Fluid (ACSF): Make Fresh Daily

125 mM NaCl, 2.5 mM KCl, 1.2 mM NaH_2PO_4, 1.2 mM $MgCl_2$, 2.4 mM $CaCl_2$, 26 mM $NaHCO_3$, and 11 mM glucose.

Note: ACSF can be prepared as a 10× stock solution, excluding $NaHCO_3$ and glucose. 1× solutions are made daily from the 10× stock, with fresh $NaHCO_3$ and glucose added.

6.2 Sucrose-Supplemented ACSF (SACSF): Make Fresh Daily

250 mM sucrose, 2.5 mM KCl, 1.2 mM NaH_2PO_4, 1.2 mM $MgCl_2$, 2.4 mM $CaCl_2$, 26 mM $NaHCO_3$, and 11 mM glucose.

Note: SACSF can be prepared as a 10× stock solution, excluding sucrose, $NaHCO_3$ and glucose. 1× solutions are made daily from the 10× stock, with fresh sucrose, $NaHCO_3$ and glucose added.

6.3 Sulfo-N-hydroxysuccinyl-SS-Biotin (Sulfo-NHS-SS-biotin, Pierce Chemical Company)

Stock solutions should be 200 mg/ml in DMSO and are resistant to multiple freeze–thaw cycles. Aliquots are stored at −20 °C. The succinyl ester is rapidly hydrolyzed in aqueous solution, so working solutions should be prepared immediately prior to applying to slices.

6.4 Slice Quench Solution

ACSF supplemented with 100 mM glycine.

6.5 RIPA Lysis Buffer

10 mM Tris, pH 7.4, 150 mM NaCl, 1.0 mM EDTA, 1% Triton-X-100, 0.1% SDS, 1% Na deoxycholate.

6.6 RIPA with Protease Inhibitors (RIPA/PI): Make Fresh Daily

RIPA supplemented with 1 μM leupeptin, 1 μM pepstatin, 1 μM aprotinin, and 1 mM phenylmethyl sulfonyl fluoride.

7 Protocol

1. Prepare Brain Slices.
   (a) Make fresh SACSF and ACSF.
   (b) Chill SACSF on ice in a small beaker to hold mouse brain.
   (c) Oxygen-saturate ACSF and SACSF by bubbling with 95%/5% O_2/CO_2, 20 min on ice.
   (d) For optimal tissue viability, P30-38 mice should be used. Sacrifice animals by cervical dislocation and decapitation, and rapidly remove brains into chilled SACSF.
   (e) Using a Vibratome, prepare 300 μm sections of the brain region of interest, and allow slices to recover for 40 min, 31 °C in oxygenated ACSF, bubbling gently. If desired, slices can be further dissected prior to recovery to enrich for particular brain regions, or to separate right and left section hemispheres to use as control and experimental slices, respectively.

2. Drug Treatment (if appropriate).
   (a) Following recovery, transfer individual slices to separate microcentrifuge tubes and wash three times in warmed (37 °C), oxygenated ACSF bubbling constantly with 95%/5% O_2/CO_2.
3. Add test compounds and incubate with continuous oxygenation.
   (a) To achieve this, we spike in a 1/10th volume of 10× drug concentrate, and gently mix by inversion.
   (b) In order to oxygenate samples, insert an 18 g needle, attached to manifold tubing, through the microcentrifuge tube lid and bubble gently with 95%/5% O_2/CO_2.
   (c) Tubes are held in a vented rack and shaken gently in a shaking water bath at the desired temperature.
4. Following drug treatment, rapidly chill slices by washing three times in ice-cold ACSF. For buffer changes between washes, allow slices to settle with gravity and gently aspirate off buffer.
5. Biotinylate surface proteins.
   (a) Prepare 1.0 mg/mL sulfo-NHS-SS-biotin in ice-cold ACSF immediately prior to labeling.
   (b) Add sulfo-NHS-SS-biotin to slices and incubate slices on ice, for 45 min.
   (c) Wash slices three times quickly with ice-cold ACSF, then incubate for 10 min in ice-cold ACSF on ice.
   (d) Wash slices three times with ice-cold slice quench buffer and incubate with slice quench buffer two times, for 25 min, on ice to quench free sulfo-NHS-SS-biotin.
6. Prepare Tissue Lysates.
   (a) Wash slices three times in ice-cold ACSF.
   (b) Gently pellet slice by centrifuging at 200 × *g*, for 1 min and carefully aspirate remaining ACSF.
   (c) Add 400 μl ice-cold RIPA/PI and break up tissue by pipetting up and down once through a P200 pipette.
   (d) Transfer dissociated slice/RIPA to a fresh tube and incubate for 30 min, at 4 °C, rotating, to complete lysis.
   (e) Pellet cellular debris by centrifuging, 18,000 × *g*, 15 min, 4 °C.
   (f) Determine lysate protein concentrations using the BCA protein assay, with BSA as a standard.
7. Isolate Biotinylated Proteins.

(a) Optimize bead–lysate ratio.

Note: Transporter expression levels can vary widely across brain regions and it is essential to quantitatively recover all of the biotinylated proteins from a given amount of tissue lysate in order to reliably determine whether significant changes in transporter surface expression have occurred. Thus, the optimal bead–lysate ratio for a particular transporter/brain region must be empirically prior to initiating any new slice biotinylation study. For DAT isolation in striatum, we use a ratio of 40 μl streptavidin beads for 20 μg striatal lysate

(b) Incubate 25 μl streptavidin agarose beads with increasing amounts of tissue lysate (approximate range 25–200 μg), and then proceed as described below for binding, washing, and elution steps.

(c) Quantify the resulting immunoreactive bands and choose a bead–lysate ratio in the linear range of binding that will permit accurate quantification of either increased or decreased transporter surface expression.

8. Prepare Streptavidin-Agarose Beads.

(a) Determine total bead volume needed for all samples. Prepare a sufficient bead volume for that amount plus one extra (i.e., four samples at 25 μl/sample = 100 μl beads + one extra = 125 μl beads).

(b) Vortex streptavidin bead stock and pipette out desired volume. If using a P200 pipette, cut off the end of the tip to prevent blockage or bead damage.

(c) Wash beads three times in 0.5–1.0 ml RIPA/PI to remove preservatives, vortexing after each RIPA addition and collecting beads between washes by centrifuging at 18,000 × *g*, for 1 min, at room temperature.

(d) Aspirate off buffer between washes using a glass Pasteur pipette attached to a vacuum flask. We find that attaching a plastic P200 tip to the end of the glass Pasteur pipette provides finer control when aspirating. It is not necessary to remove absolutely all the buffer for the first two washes, as it risks aspirating beads into the pipette. After the final wash, remove as much excess buffer as possible without aspirating the beads.

(e) Add RIPA/PI to bring beads back to their original volume, pipetting up and down to resuspend. Avoid vortexing at this point, as beads will stick to the tube wall.

(f) Aliquot beads into microcentrifuge tubes using a P200 with the tip cut off. Pipette up and down several times between sampling to assure beads remain evenly suspended and dispersed among the tubes.

9. Bind biotinylated proteins to streptavidin beads.
   (a) Distribute cell lysates to the tubes containing the agarose beads. For experiments where multiple samples are being compared, make sure to use the same amount of protein for each sample for accurate comparisons.
   (b) Add additional RIPA/PI to bring to samples to a 200 μl minimal volume. This assures that samples will mix adequately during the incubation and also unifies the protein concentrations across samples.
   (c) Place tubes in a tube rotator and mix overnight at 4 °C.
   (d) In separate tubes, dispense an equivalent of the total lysate volume used for each sample. Alternatively, if the sample volumes are high, a fraction of total lysate volume can be reserved instead, to accommodate maximal load volumes on SDS-PAGE gels. These will be used to normalize the surface expression to the total protein amount.
   (e) Add either an equal volume of 2× or 1/5 volume of 6× SDS-PAGE sample buffer and store at −20 °C until samples are analyzed.
10. Elute and analyze samples.
   (a) Pellet beads by centrifuging at 18,000 × *g*, for 2 min, at room temperature.
   (b) Aspirate supernatant and wash beads three times with 0.75 ml RIPA. To minimize bead loss, leave a small head volume of buffer above beads between washes. After the final wash, remove as much RIPA as possible, without disrupting the bead pellet.
   (c) Elute biotinylated proteins from streptavidin beads by reducing the disulfide linkage. Add 25 μl 2× SDS-PAGE reducing sample buffer, vortex well, and rotate samples for 30 min, at room temperature.

   Note: Polytopic transporters have a high tendency to aggregate when boiled in SDS-PAGE sample buffer, severely impairing their electrophoretic mobility. Thus, we suggest avoiding heating samples and eluting slowly at room temperature. If boiling is absolutely necessary (e.g., if another protein assessed in parallel requires boiling), sample buffer can be supplemented with 2 M urea (final concentration) to minimize aggregation. While effective in reducing aggregation in some cases, urea often compromises band appearances.
11. Analyze samples by immunoblot.
12. Thaw total lysate samples and rotate in parallel with bead samples, for 30 min, at room temperature.
13. Separate proteins on SDS-PAGE gels.

14. Indentify protein(s) of interest by immunoblotting.
    (a) Be certain that bands are detected in the linear range of detection for proper quantification.
    (b) To assure that the biotinylation reagent has not gained access to intracellular proteins via damaged/compromised cells, we recommend immunoblotting in parallel for an intracellular protein specific to the cell type being investigated.
    (c) Quantify band densities and calculate relative transporter surface density as a % of the total transporter expression level.

8 Representative Result

PKC-regulated dopamine transporter (DAT) endocytic trafficking has long been described in a variety of neuronal and non-neuronal cell lines [20–22, 25, 26]. In contrast, several reports indicate a lack of PKC-stimulated DAT internalization in cultured dopaminergic neurons derived from rodent ventral midbrain [27–29]. We used slice biotinylation to ask whether PKC activation drives DAT internalization in adult mouse striatal neurons. As seen in Fig. 2, we reproducibly detect robust DAT surface levels in mouse striatum (81.4 ± 5.8% of total DAT) with minimal intracellular protein biotinylation, as determined for tyrosine hydroxylase in parallel (TH; 1.6 ± 0.4% of total TH). PKC activation with 1 μM PMA, 30 min, 37 °C significantly decreased surface DAT to 60.8 ± 5.2% total DAT, which translates to ~30% surface loss. These results are in agreement with those of Cremona and associates [30] and demonstrate (1) that DAT undergoes PKC-stimulated internalization in native dopaminergic neurons and (2) that transporter trafficking can be reliably measured in adult neurons using brain slice biotinylation.

8.1 Measuring Transporter Internalization Rates by Reversible Biotinylation

Steady state biotinylation measures changes in transporter surface expression in response to a given stimulus. Acute changes in transporter surface expression may be due to accelerated internalization or dampened endocytic recycling back to the plasma membrane. In order to mechanistically distinguish between these two possibilities, it is essential to measure both internalization and recycling rates. Reversible biotinylation measures the amount of biotinylated protein that internalizes over a given time period, thereby allowing for rate measurements. This approach capitalizes on the ability to remove surface biotin moieties via reducing the disulfide linker that couples the biotin to the reactive NHS ester group. Transporter surface pools are first biotinylated under trafficking restrictive temperatures (0 °C), and then are permitted to internalize by shifting to trafficking permissive conditions (37 °C) for short time periods. Cells are subsequently treated with membrane impermeant reductants that strip residual surface biotinylation by reducing the disulfide linker, thereby liberating

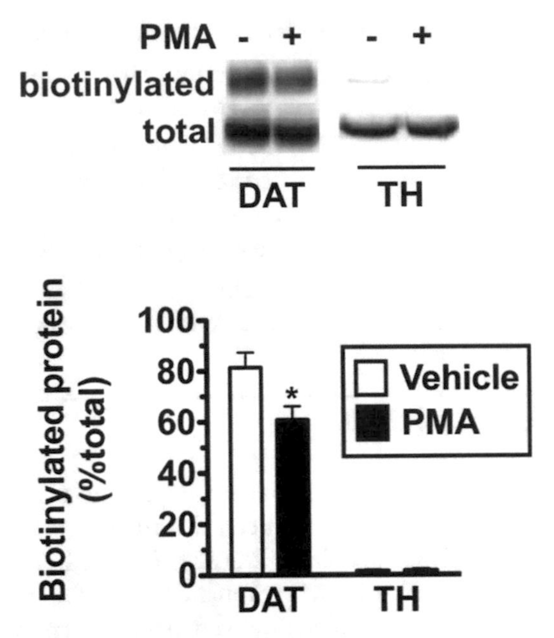

Fig. 2 PKC activation decreases DAT surface levels in adult mouse striatal neurons. *Brain slice biotinylation:* Striatal slices were treated ±1 μM PMA, 30 min, 37 °C and surface proteins were biotinylated and isolated as described in *Protocol*. Immunoblots were probed for DAT and tyrosine hydroxylase (TH) as an intracellular reporter. *Top:* Representative immunoblots showing biotinylated and total protein detected following the indicated treatments. *Bottom*: Averaged data. Data are expressed as biotinylated protein as a %total protein in sample ± S.E.M. Note lack of biotinylated TH, indicating that biotinylation was restricted to surface proteins in dopaminergic neurons. *Significantly different from vehicle control, $p < 0.03$, Student's t test, $n = 6$

the biotin moieties. Internalized, biotinylated proteins are protected from the strip, are isolated and compared to the original surface pool to determine the rate of internalization.

Note: This protocol describes a single timepoint internalization measurement that we have empirically determined to be optimal for the dopamine transporter in cell lines [21, 22]. However, in order to accurately use a single timepoint measurement, it is critical to know a priori how rapidly a given protein returns to the cell surface after internalizing. Over time, the amount of biotinylated protein that is protected from reduction will plateau as proteins undergo endocytic recycling. Thus, if too long an internalization incubation time is used, biotinylated proteins may have reequilibrated between the surface and endosomal pool and rate differences between control and experimental samples will not be discernable. Therefore, we highly recommend beginning these studies by performing an internalization timecourse to fully understand the internalization profile of a given protein prior to choosing a single timepoint to measure. A single timepoint should ideally be chosen in the linear portion of the internalization curve, so that both increases and decreases in the rate can be accurately measured.

8.2 Required Reagents

PBS2+

Phosphate-buffered saline, pH 7.4, supplemented with 1.0 mM $MgCl_2$, 0.1 mM $CaCl_2$.

Quench Solution

PBS^{2+} supplemented with 100 mM glycine.

RIPA Lysis Buffer

10 mM Tris, pH 7.4, 150 mM NaCl, 1.0 mM EDTA, 1% Triton-X-100, 0.1% SDS, 1% Na deoxycholate.

RIPA with Protease Inhibitors (RIPA/PI)—Make fresh daily

RIPA supplemented with 1 μM leupeptin, 1 μM pepstatin, 1 μM aprotinin, and 1 mM phenylmethyl sulfonyl fluoride.

"*Happy*" *PBS*—Make fresh daily

PBS^{2+}, 0.18% (w/v) glucose, 0.2% IgG-free, protease-free BSA (Sigma-Aldrich #A3059).

NT Buffer

150 mM NaCl, 1.0 mM EDTA, 0.2% BSA, 20 mM Tris, pH 8.6.

Tris(2-carboxyethyl)phosphine (TCEP, Pierce Chemical Company)

500 mM TCEP stock solutions should be made in 50 mM Tris, pH 8.6 and further adjusted to pH 8.6. Note that TCEP is highly acidic and requires pH adjustment even under strong buffering conditions. Store aliquoted stock solutions at –20 °C in light protected condition.

Sulfo-N-hydroxysuccinyl-SS-Biotin (sulfo-NHS-SS-biotin, Pierce Chemical Company)

Stock solutions should be 200 mg/ml in DMSO and are resistant to multiple freeze–thaw cycles. Aliquots are stored at –20 °C. The succinyl ester is rapidly hydrolyzed in aqueous solution, so working solutions should be prepared immediately prior to applying to cells.

Streptavidin Agarose (Pierce Chemical Company)

Provided as a 50% slurry. Mix well prior to sampling.

9 Protocol

1. Cell plating.
   (a) For each transporter being investigated, plate (one) well as a zero time point, (one) well as a stripping control, and (one) well for each internalization conditions being investigated.
   (b) Zero time point and strip control wells should be plated in a separate plate from internalized samples, as they will

remain at 0 °C, whereas experimental samples will be warmed to 37 °C.

2. Biotinylate surface proteins.
3. Shift cells to trafficking restrictive temperature.
   (a) Place on an ice bath at 4 °C.
   (b) Wash cells 4 × 2 ml w/ice-cold PBS^{2+}.
   (c) Drain all residual buffer by tilting plates and aspirating.
4. Make 2.5 mg/ml sulfo-NHS-SS-biotin in PBS^{2+}.
   (a) Prepare a sufficient amount to evenly coat the bottom of each well.
   (b) 0.75 ml/well is sufficient for one well of a 6-well culture plate.
5. Add 0.75 ml sulfo-NHS-SS-biotin per well and incubate for 15 min, at 4 °C on ice, shaking vigorously to assure complete coverage of the well.
6. Make another batch of 2.5 mg/ml sulfo-NHS-SS-biotin in PBS^{2+}, aspirate and discard old solution and replace wells w/ fresh solution. Incubate for 15 min, at 4 °C on ice, shaking.
7. Wash rapidly three times with 2.0 ml/well ice-cold quench solution.
8. Quench residual reactive NHS groups by incubating twice in 1–2 ml quench solution/well × 15 min, 4 °C on ice, shaking.
9. Initiate internalization.
   (a) Control plates with time zero and strip control wells remain on ice bath at 4 °C on ice.
   (b) Transfer experimental samples to room temperature bench top and warm experimental samples by washing 3 × 2 ml with prewarmed (37 °C) Happy PBS containing any drugs of interest.
   (c) Incubate is same solutions for 10 min, 37 °C, or use separate plates and warm to various times to obtain a time-course of internalization.
10. Strip residual surface biotin.
    (a) Prepare TCEP reductant.
    (b) Dilute TCEP stocks to 50 mM final concentration in ice-cold NT Buffer. Prepare a sufficient amount for 1.0 ml/well for each internalized and strip well.
    (c) Keep on ice until internalization is completed.
11. Strip.
    (a) Rapidly cool internalized cells by placing on ice bath, moving to the cold room, and washing 3 × 2.0 ml with ice-cold NT buffer.

(b) Also wash strip control wells 3×2.0 ml with ice-cold NT buffer.

(c) Add 1.0 ml/well of 50 mM TCEP solution to internalized and strip control well.

(d) Incubate for 15 min, at 4 °C on ice, shaking.

(e) Prepare another batch of 50 mM TCEP/NT buffer.

(f) Remove first TCEP solution from cells and replace with fresh TCEP solution.

(g) Incubate for 15 min, at 4 °C on ice, shaking.

(h) Wash cells 3×2.0 ml with NT buffer.

(i) Wash all wells (including time zero control) 3×2.0 ml PBS^{2+} to remove all residual NT buffer.

Note: Residual BSA in NT buffer could artifactually enhance apparent protein concentrations after cell lysis.

12. Lyse cells, isolate biotinylated protein and analyze.

(a) Lyse cells and determine lysate protein content as described for steady state biotinylation.

(b) Isolate biotinylated proteins from each sample as described for steady state biotinylation. Note that no "total" samples are necessary, as all internalization will be normalized to time zero control surface signals for each sample.

(c) Immunoblot total, strip control and experimental samples for transporter of interest. Calculate strip efficiency, comparing strip samples to time zero surface levels. Ideally, strip efficiencies should be >95%, otherwise it is difficult to assure that experimental signals are due to internalization, as opposed to residual surface signal. Quantify internalization as fraction internalized in a given time period, as compared to the total surface pool at time zero.

10 Representative Results

SLC6 transporter carboxy termini encode multiple signals required for ER export [31–38], surface targeting [39–41] and PKC-stimulated internalization [19–21]. Moreover, the DAT carboxy terminus is a binding locus for multiple proteins involved in regulating DAT trafficking and substrate efflux [39, 42–44]. In a previous report [19], we used site-directed mutagenesis and reversible biotinylation to ask whether residues in the DAT 587–596 carboxy terminal span were required for PKC-stimulated DAT internalization. As seen in Fig. 3, wildtype DAT constitutively internalized at 28.3±5.3% over 10 min in PC12 cells and PKC activation with 1 μM PMA significantly accelerated DAT internalization rates to 57.2±6.0% over 10 min. When DAT carboxy terminal residues

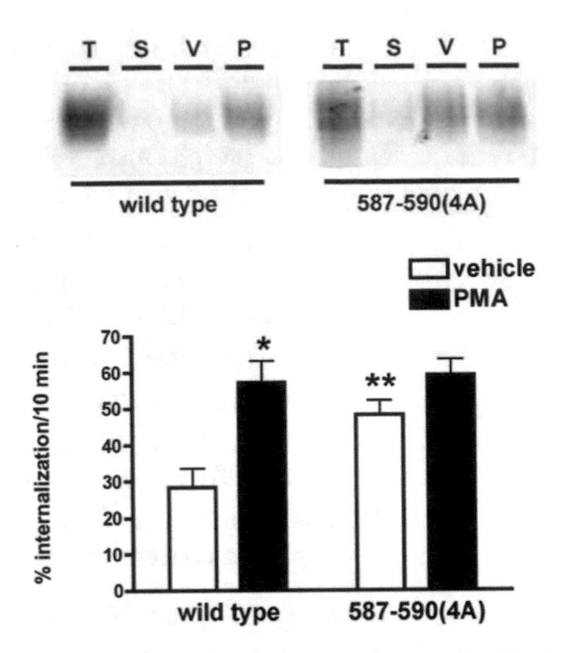

Fig. 3 DAT carboxy terminal residues control basal and PKC-stimulated DAT endocytic rates. *Internalization rates measured by reversible biotinylation.* PC12 cells stably expressing either wildtype or 587–590(4A) DAT were biotinylated (0 °C) and internalization rates were measured ±1 μM PMA, 10 min, 37 °C as described in *Protocol. Top:* Representative blots showing total DAT at $t=0$ (T), strip control (S), and internalized samples under vehicle (V) or PMA (P) treatment. *Bottom*: Averaged data. Data are expressed as %total surface pool internalized/10 min ± S.E.M. as compared to $t=0$ control samples. *Adapted from* [15]

587–590 were mutated to serines (DAT 587–590(4A)), we observed a significant increase in DAT constitutive endocytosis as compared to wildtype DAT, to 48.1 ± 4.1 % over 10 min. Moreover, PKC activation did not further increase DAT 587–590(4A) mutant internalization rates. These results demonstrate that a negative regulatory "endocytic brake" dampens DAT internalization rates and is released upon PKC activation. Sorkina and associates also reported that the DAT amino terminus contributes to the DAT endocytic brake [23]. Moreover, a coding variant in the DAT carboxy terminus of an ADHD patient was recently reported, that is lacking in endocytic braking [14]. Thus, the ability to measure transporter internalization rates can shed light on basic mechanisms of transporter regulation, as well as offer insight into molecular basis of neuropsychiatric disorders.

10.1 Measuring Transporter Endocytic Recycling Rates by Temperature Shift Biotinylation

This approach measures transporter insertion rates into the plasma membrane over time. The initial surface pool at time zero is biotinylated at trafficking restrictive temperatures (0 °C). Then, cells are shifted to the trafficking permissive temperature (37 °C) and are biotinylated so that newly inserted proteins will be labeled, resulting in an increase in biotinylation over baseline values. An important consideration with this method is that the reducing environment of the cytoplasm can be sufficient to reduce the disulfide linker coupling the biotin to the reactive NHS group, resulting in a loss of biotinylation over time rather than an increase. Thus, it is important to include control samples at each timepoint that are warmed in parallel in the absence of the biotinylating reagent, to determine if biotinylation losses occur at the trafficking permissive temperature. If disulfide reduction poses a significant obstacle in measuring the recycling rate, sulfo-NHS-biotin, rather than sulfo-NHS-SS-biotin, can alternatively be used. In this case, biotinylated proteins cannot be eluted from streptavidin beads by reduction, since there is no disulfide linker present. Thus, biotinylated proteins must be isolated with monomeric avidin beads. This form of avidin has a significantly lower binding affinity than tetrameric streptavidin (10^{-9} M vs. 10^{-14} M), and is easily eluted with SDS-PAGE sample buffer containing 2 mM free biotin.

11 Required Reagents

PBS2+

Phosphate-buffered saline, pH 7.4, supplemented with 1.0 mM $MgCl_2$, 0.1 mM $CaCl_2$.

PBS2+/g

PBS^{2+} supplemented with 0.18% glucose.

Quench Solution

PBS^{2+} supplemented with 100 mM glycine.

RIPA Lysis Buffer

10 mM Tris, pH 7.4, 150 mM NaCl, 1.0 mM EDTA, 1% Triton-X-100, 0.1% SDS, 1% Na deoxycholate.

RIPA with Protease Inhibitors (RIPA/PI)—Make fresh daily

RIPA supplemented with 1 μM leupeptin, 1 μM pepstatin, 1 μM aprotinin, and 1 mM phenylmethyl sulfonyl fluoride.

Sulfo-N-hydroxysuccinyl-SS-Biotin (sulfo-NHS-SS-biotin, Pierce Chemical Company)

Stock solutions should be 200 mg/ml in DMSO and are resistant to multiple freeze–thaw cycles. Aliquots are stored at −20 °C. The succinyl ester is rapidly hydrolyzed in aqueous solution, so working solutions should be prepared immediately prior to applying to cells.

Streptavidin Agarose (Pierce Chemical Company)

Provided as a 50% slurry. Mix well prior to sampling.

12 Protocol

1. Cell plating.
2. For each condition being investigated, plate (1) well as 4 °C control and (1) well for each timepoint being measured.
3. The 4 °C controls should be on a separate plate kept at 4 °C after initial labeling. Separate plates should be prepared for each 37 °C timepoint, and each timepoint should include a control well for biotinylation loss, as well as control and experimental sample wells. Representative timepoints of 2, 5, 10, 15, 30, and 45 min as usually sufficient to yield reliable results.
4. Biotinylate all surface proteins.
   (a) Shift cells to trafficking restrictive temperature.
      - Place on an ice bath at 4 °C.
      - Wash cells 4 × 2 ml w/ice-cold PBS2+.
      - Drain all residual buffer by tilting plates and aspirating.
5. Make 1.0 mg/ml sulfo-NHS-SS-biotin in PBS^{2+}.
   (a) Prepare a sufficient amount to evenly coat the bottom of each well.
   (b) 0.75 ml/well is sufficient for one well of a 6-well culture plate.
6. Add 0.75 ml sulfo-NHS-SS-biotin per well and incubate for 15 min, at 4 °C on ice, shaking vigorously to assure complete coverage of the well.
7. Make another batch of 1.0 mg/ml sulfo-NHS-SS-biotin in PBS^{2+}, aspirate and discard old solution and replace wells w/ fresh solution. Incubate for 15 min, at 4 °C on ice, shaking.
8. Wash rapidly three times with 2.0 ml/well ice-cold quench solution.
9. Quench residual reactive NHS groups by incubating twice in 1-2 ml quench solution/well for 15 min, at 4 °C on ice, shaking.
10. Initiate Recycling.
    (a) Leave 4 °C control plate at 4 °C during 37 °C timecourse. Warm up timepoint plates one at a time, starting with longest timepoint and ending with shortest.
    (b) Transfer experimental samples to room temperature bench top and warm experimental samples by washing rapidly 3 × 2 ml with prewarmed (37 °C) PBS^{2+}/g containing any drugs of interest.

Note: Cells are warmed in PBS^{2+}/g, rather than Happy PBS, to prevent BSA from competing for the biotinylation reagent.

(c) Prepare a 1.0 mg/ml sulfo-NHS-SS-biotin solution in prewarmed PBS^{2+} ± drugs of interest and add to cells.

(d) Incubate cells at 37 °C for appropriate time. Repeat for other timepoints, working backwards from longest to shortest timepoint, such that all timepoints will be completed simultaneously.

11. Stop biotinylation.

(a) Wash cells rapidly 3 × 2 ml with ice-cold PBS^{2+}.

(b) Quench residual reactive NHS groups by incubating twice in 1–2 ml quench solution/well × 15 min, 4 °C on ice, shaking.

12. Lyse cells, isolate biotinylated protein and analyze.

(a) Lyse all cells in RIPA/PI as described for steady state biotinylation (above) and determine lysate protein content.

(b) Isolate biotinylated protein as described for *steady state biotinylation* (above).

(c) In separate tubes, dispense an equivalent of the total lysate volume used for each sample. Alternatively, if the sample volumes are high, a fraction of total lysate volume can be reserved instead, to accommodate maximal load volumes on SDS-PAGE gels. These will be used to normalize the surface expression to the total protein amount.

(d) Immunoblot for transporter of interest.

(e) Calculate the %total protein biotinylated at each timepoint. An increase in biotinylation that plateaus should be observed over time.

(f) Data points can be normalized to the maximal biotinylated signal to account for experimental variability in absolute biotinylation levels.

(g) Data points are fit to a first order exponential equation and a rate is derived from the fit.

13 Representative Results

AMPH exposure causes a loss in surface DAT that is dependent upon CamKII and Akt activation [45, 46]. Given Akt's established role in recruiting the insulin-sensitive glucose transporter, GLUT4, to the plasma membrane [47], we asked whether dampened recycling contributes to AMPH-mediated DAT surface losses. As seen in Fig. 4a, b, DAT robustly recycled to the plasma membrane with an

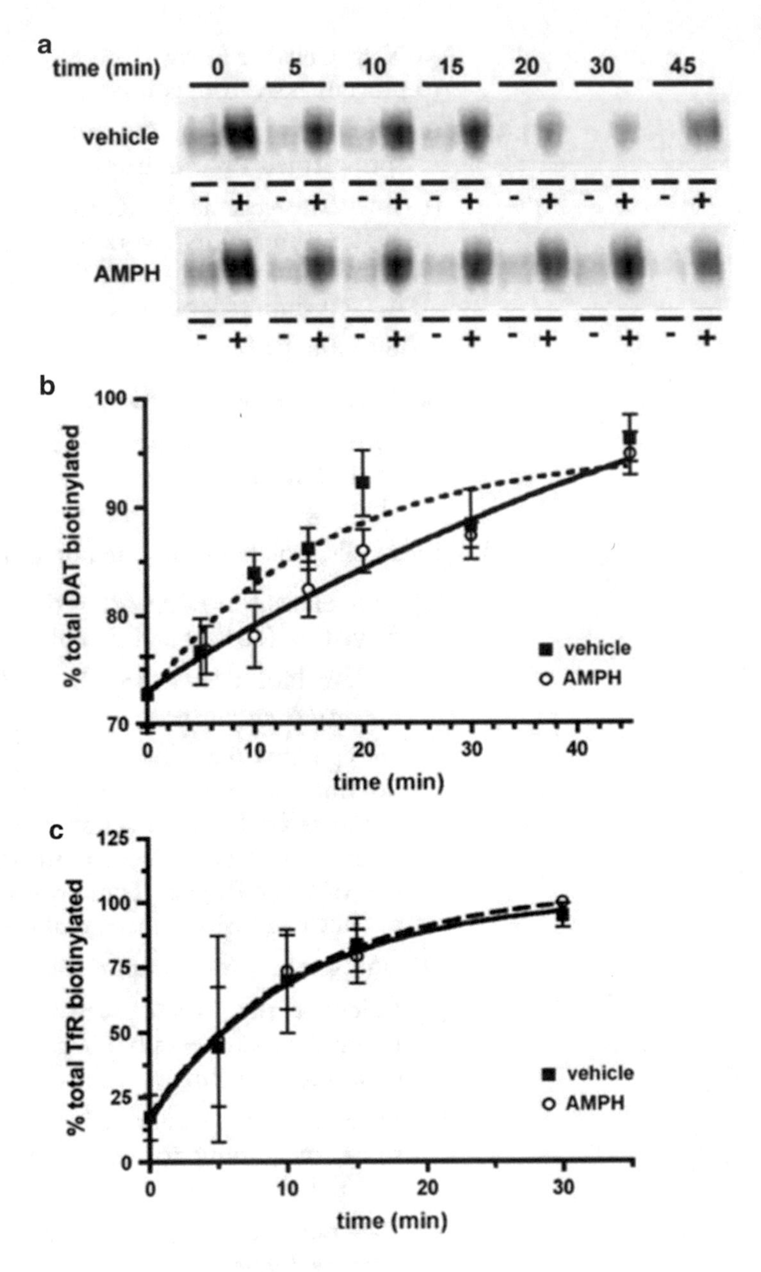

Fig. 4 AMPH treatment decreases DAT endocytic recycling rates. *Recycling* Assay. PC12 cells stably expressing DAT were biotinylated first at 0 °C, followed by biotinylation at 37 °C ±5 μM AMPH for the indicated times. Recycling rates were measured as described in *Protocol.* (**a**) Representative DAT blots showing biotinylated (+) and non-biotinylated (–) cell fractions recovered following the indicated treatments. (**b**) Average DAT data expressed at % total DAT biotinylated ± S.E.M. Data points were fit to a single exponential and rates derived using GraphPad Prism. (**c**) Average TfR data expressed as %total TfR biotinylated. Immunoblots were stripped and reprobed for the TfR. *Adapted from* [15]

average intracellular half-life of 11.2 min (0.062 min^{-1}). Treatment with 5 μM AMPH significantly decreased DAT endocytic recycling to 0.015 min^{-1}, which correlates to an intracellular half-life of 44.9 min. AMPH-mediated decreases in DAT recycling were not due to global effects on endocytic trafficking, as transferrin receptor recycling rates were unaffected by AMPH exposure (Fig. 4c).

14 Cellular Imaging Approaches

Biochemical approaches have broad utility for measuring transporter trafficking across relatively homogeneous cell populations. However, they are less reliable in instances where co-transfection/gene manipulation is required, such as co-expression of a transporter and a dominant negative or constitutively active enzyme, or with a silencing RNA. These experimental conditions often present mixed cell populations that express transporter, but not always another gene or siRNA of interest. Therefore, trafficking measurements across the entire cell population will be skewed by cells that are not co-transfected and may therefore underestimate the impact that a specific gene or cofactor has on transporter trafficking. In these instances, a single cell imaging approach is preferred, as cells can be independently assessed, with the assurance that the cell being examined expresses both the transporter and modifying gene or siRNA. This approach requires either an antibody that recognizes an extracellular epitope in the native transporter, or a transporter with an engineered extracellular epitope tag. Antibody is bound to target surface transporters in trafficking restrictive temperatures and is subsequently allowed to internalize by warming cells to trafficking permissive temperatures under control vs. experimental conditions. Residual (non-internalized) surface antibody is blocked with unconjugated secondary antibody so that only internalized antibody will be detected. Cells are then permeabilized and internalized antibody is detected with fluorophore-conjugated secondary antibody. While this method can be used to reliably determine whether or not internalization is altered, we caution against using it to track the endocytic fate of internalized proteins. Recent studies indicate that antibody binding to protein extracellular domains can dramatically alter the endocytic fate of a protein, most frequently resulting in artifactual targeting to degradative pathways [48, 49].

15 Single cell internalization assays

15.1 Reauired Regents

PBS

Phosphate-buffered saline, pH 7.4.

"Happy" PBS—Make fresh daily

PBS^{2+}, 0.18% (w/v) glucose, 0.2% IgG-free, protease-free BSA (Sigma-Aldrich #A3059).

Non-permeabilizing block (NPB)

5% Normal Goat Serum, 1% IgG Free/Protease Free BSA in PBS.

Permeabilizing block (PB)

5% Normal Goat Serum, 1% IgG Free/Protease Free BSA, 0.2% saponin in PBS.

16 Protocol

16.1 Cell Plating and Transfection (If Appropriate)

1. Plate cells on glass coverslips at a density compatible with transfection (if appropriate). It is optimal to plate cells at the minimum density possible, to optimize the ability to image single cells in isolation.
2. Plate sufficient wells for experimental and control samples. On a separate plate, seed cells for blocking (to assure surface block with unconjugated secondary antibody) and total surface staining (time zero) controls. These must be seeded on a separate plate, as they will not be warmed to the trafficking permissive temperature.
3. Transfect cells, if necessary, 1 day after plating. Perform assay at appropriate post-transfection timepoint

16.2 Label Surface Proteins with Antibody

Note: Optimally, antibody binding should be performed at a trafficking restrictive temperature (e.g., 4 °C), so that internalization can be measured starting at a true zero timepoint. However, there are instances where an antibody fails to bind its epitope in a native protein under low temperature conditions. In those instances, the investigator must empirically determine the optimal antibody binding temperature that allows for maximal antibody binding with minimal constitutive transporter internalization. In the example below, we give a detailed protocol for binding an extracellular HA epitope engineered into the DAT, which can only bind at temperatures greater than 18 °C.

1. Wash cells three times/well with Happy PBS.
2. Incubate cells with 2.0 μg/ml mouse anti-HA (clone HA.11, Covance) in Happy PBS, for 30 min, at room temperature.

16.3 Initiate Internalization and Process Control Cells

1. For control cells:
   (a) Fix cells in 4% paraformaldehyde prepared in PBS, for 10 min, room temperature.
   (b) Hold fix cells until experimental samples are fixed, then process in parallel.

16.4 For Experimental Cells

1. Wash cells rapidly three times in prewarmed Happy PBS containing any experimental test solutions.
2. Incubate cells in the appropriate prewarmed experimental solutions for 15 min, at 37 °C to permit internalization.
3. Wash cells rapidly three times with ice-cold PBS to stop internalization.
4. Fix cells in 4% paraformaldehyde prepared in PBS, for 10 min, at room temperature.

16.5 Detect Internalized Antibodies

1. For negative control cells (blocking control):
   (a) Block surface antibodies by incubating cells with NPB containing 20 μg/ml unconjugated goat anti-mouse secondary antibody, for 45 min, at room temperature.
   (b) Wash cells three times, 5 min each with PBS.

16.6 For Zero Timepoint Surface Control (Non-permeabilized Cells)

1. Detect surface antibody binding by incubating cells with NPB containing goat anti-mouse secondary antibody conjugated to a fluorophore (1:5000) compatible with the investigator's imaging system, 45 min, room temperature.
2. Wash cells three times, 5 min each with PBS.

16.7 For Experimental Cells

1. Block residual surface antibodies with NPB containing 20 μg/ml unconjugated goat anti-mouse antibody, for 45 min, at room temperature.
2. Wash cells three times with PBS and permeabilize cells with PB, for 30 min, at room temperature.
3. If appropriate, stain cells with primary antibodies directed against any marker proteins (e.g., transfection markers that assure co-transfection). Wash cells three times with PBS, 5 min each.
4. Probe cells for internalized primary antibody with 1:5000 goat anti-mouse secondary antibody and any fluorophore-coupled secondary antibodies to detect marker proteins (if appropriate), 45 min, at room temperature.
5. Wash cells three times, 5 min each with PBS.

16.8 Dry and Mount Coverslips

1. Dry coverslips for 20-30 min, 37 °C.
2. Mount onto slides with anti-quench mounting medium.

16.9 Analyze Internalization

1. Image individual cells either with confocal or wide-field microscopy. If using wide-field microscopy, capture z-stacks and deconvolved image stack to assure localization of putative intracellular signals.

(a) Optimize image capture conditions for experimental samples first. Use identical image capture conditions (exposure times) for blocking controls, so that background fluorescence signal can be quantified.

(b) Internalization can be measured in two ways:

- *Score cells:* The investigator capturing the images assigns code values to the images. A second investigator blindly scores the cells for internalization by the obvious appearance of puncta as compared to negative control cells. The code is then revealed, cells are assembled into their respective experimental groups, and the %cells demonstrating internalization is calculated.
- *Quantify fluorescence over background*: Using an image processing program, each cell is masked and total fluorescence quantified. Fluorescence intensities of experimental cells are calculated as fold increase over background signals in negative control (blocked) cells. If this approach is used, it is imperative that capture conditions (exposure times) are identical for all cells, as described in Sect. 6.1.1.

17 Representative Results

We recently reported that the neuronal GTPase, Rin, binds to the DAT carboxy terminus in a PKC-regulated manner [44]. We used Rin-directed shRNA to test whether Rin was required for PKC-stimulated DAT internalization in the human dopaminergic neuroblastoma cell line SK-N-MC. We were unable to achieve high co-transfection efficiencies with DAT and shRNA plasmids, such that we could not be certain that all DAT-transfected cells also expressed the Rin-directed shRNA. Therefore, we chose a single cell internalization approach with cell scoring to determine whether Rin knockdown impaired PKC-stimulated DAT internalization. As seen in Fig. 5, cells co-transfected with DAT and either vector or scrambled shRNA exhibited robust intracellular puncta in response to PKC activation in $79.8 \pm 2.9\%$ and $73.0 \pm 11.0\%$ of cells imaged, respectively. In contrast, significantly fewer cells co-transfected with DAT and Rin228 shRNA exhibited discernable intracellular puncta ($37.2 \pm 8.9\%$ cells, $p < 0.05$, one-way ANOVA with Dunnett's post hoc analysis). These results clearly demonstrate the utility of the single cell internalization assay in experimental conditions with mixed cell populations.

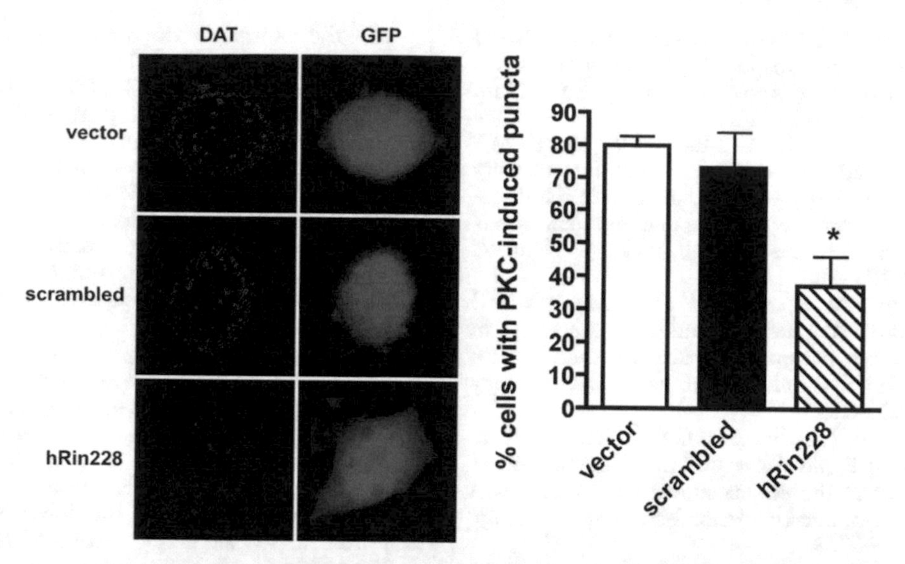

Fig. 5 Rin GTPase is required for PKC-mediated DAT internalization. *Antibody internalization assay*. SK-N-MC cells were co-transfected with DAT encoding an HA epitope in the second extracellular loop and the indicated constructs, and were assayed 72 h post-transfection. Cells were labeled with HA antibody, and DAT internalization was assessed for 15 min, 37 °C in the presence of 1 μM PMA. Cells expressing shRNA constructs were identified by GFP co-expression. *Left:* Representative images. Single planes through the cell center are shown and are representative of 30–34 cells imaged per condition over three independent experiments. *Right:* Averaged data expressed as %cells exhibiting discernable PMA-induced DAT puncta ± S.E.M. *Significantly different from vector-transfected control, $p < 0.05$ (one-way ANOVA with Dunnett's multiple comparison test; $n = 30–34$). *Adapted from* [44]

References

1. Blakely RD, Edwards RH (2012) Vesicular and plasma membrane transporters for neurotransmitters. Cold Spring Harb Perspect Biol 4(2):pii: a005595
2. Kristensen AS, Andersen J, Jorgensen TN et al (2011) SLC6 neurotransmitter transporters: structure, function, and regulation. Pharmacol Rev 63:585–640
3. Torres GE, Gainetdinov RR, Caron MG (2003) Plasma membrane monoamine transporters: structure, regulation and function. Nat Rev Neurosci 4:13–25
4. Nirenberg MJ, Chan J, Pohorille A et al (1997) The dopamine transporter: comparative ultrastructure of dopaminergic axons in limbic and motor compartments of the nucleus accumbens. J Neurosci 17:6899–6907
5. Nirenberg MJ, Chan J, Vaughan RA et al (1997) Immunogold localization of the dopamine transporter: an ultrastructural study of the rat ventral tegmental area. J Neurosci 17:4037–4044
6. Nirenberg MJ, Vaughan RA, Uhl GR et al (1996) The dopamine transporter is localized to dendritic and axonal plasma membranes of nigrostriatal dopaminergic neurons. J Neurosci 16:436–447
7. Melikian HE (2004) Neurotransmitter transporter trafficking: endocytosis, recycling, and regulation. Pharmacol Ther 104:17–27
8. Robinson MB (2002) Regulated trafficking of neurotransmitter transporters: common notes but different melodies. J Neurochem 80:1–11
9. Gulley JM, Zahniser NR (2003) Rapid regulation of dopamine transporter function by substrates, blockers and presynaptic receptor ligands. Eur J Pharmacol 479:139–152
10. Zahniser NR, Sorkin A (2004) Rapid regulation of the dopamine transporter: role in stimulant addiction? Neuropharmacology 47(Suppl 1):80–91
11. Gether U, Andersen PH, Larsson OM et al (2006) Neurotransmitter transporters: molecular function of important drug targets. Trends Pharmacol Sci 27:375–383
12. Zahniser NR, Sorkin A (2009) Trafficking of dopamine transporters in psychostimulant actions. Semin Cell Dev Biol 20:411–417

13. Barwick KE, Wright J, Al-Turki S et al (2012) Defective presynaptic choline transport underlies hereditary motor neuropathy. Am J Hum Genet 91:1103–1107
14. Sakrikar D, Mazei-Robison MS, Mergy MA et al (2012) Attention deficit/hyperactivity disorder-derived coding variation in the dopamine transporter disrupts microdomain targeting and trafficking regulation. J Neurosci 32:5385–5397
15. Boudanova E, Navaroli DM, Melikian HE (2008) Amphetamine-induced decreases in dopamine transporter surface expression are protein kinase C-independent. Neuropharmacology 54:605–612
16. Johnson LA, Furman CA, Zhang M et al (2005) Rapid delivery of the dopamine transporter to the plasmalemmal membrane upon amphetamine stimulation. Neuropharmacology 49:750–758
17. Saunders C, Ferrer JV, Shi L et al (2000) Amphetamine-induced loss of human dopamine transporter activity: an internalization-dependent and cocaine-sensitive mechanism. Proc Natl Acad Sci U S A 97:6850–6855
18. Sorkina T, Caltagarone J, Sorkin A (2013) Flotillins regulate membrane mobility of the dopamine transporter but are not required for its protein kinase C dependent endocytosis. Traffic 14:709–724
19. Boudanova E, Navaroli DM, Stevens Z et al (2008) Dopamine transporter endocytic determinants: carboxy terminal residues critical for basal and PKC-stimulated internalization. Mol Cell Neurosci 39:211–217
20. Sorkina T, Hoover BR, Zahniser NR et al (2005) Constitutive and protein kinase C-induced internalization of the dopamine transporter is mediated by a clathrin-dependent mechanism. Traffic 6:157–170
21. Holton KL, Loder MK, Melikian HE (2005) Nonclassical, distinct endocytic signals dictate constitutive and PKC-regulated neurotransmitter transporter internalization. Nat Neurosci 8:881–888
22. Loder MK, Melikian HE (2003) The dopamine transporter constitutively internalizes and recycles in a protein kinase C-regulated manner in stably transfected PC12 cell lines. J Biol Chem 278:22168–22174
23. Sorkina T, Richards TL, Rao A et al (2009) Negative regulation of dopamine transporter endocytosis by membrane-proximal N-terminal residues. J Neurosci 29:1361–1374
24. Wu S, Bellve KD, Fogarty KE et al (2015) Ack1 is a dopamine transporter endocytic brake that rescues a trafficking-dysregulated ADHD coding variant. Proc Natl Acad Sci U S A 112:15480–15485
25. Melikian HE, Buckley KM (1999) Membrane trafficking regulates the activity of the human dopamine transporter. J Neurosci 19:7699–7710
26. Daniels GM, Amara SG (1999) Regulated trafficking of the human dopamine transporter. Clathrin-mediated internalization and lysosomal degradation in response to phorbol esters. J Biol Chem 274:35794–35801
27. Sorkina T, Miranda M, Dionne KR et al (2006) RNA interference screen reveals an essential role of Nedd4-2 in dopamine transporter ubiquitination and endocytosis. J Neurosci 26:8195–8205
28. Eriksen J, Rasmussen SG, Rasmussen TN et al (2009) Visualization of dopamine transporter trafficking in live neurons by use of fluorescent cocaine analogs. J Neurosci 29:6794–6808
29. Rao A, Simmons D, Sorkin A (2011) Differential subcellular distribution of endosomal compartments and the dopamine transporter in dopaminergic neurons. Mol Cell Neurosci 46:148–158
30. Cremona ML, Matthies HJG, Pau K et al (2011) Flotillin-1 is essential for PKC-triggered endocytosis and membrane microdomain localization of DAT. Nat Neurosci 14:469–477
31. Miranda M, Sorkina T, Grammatopoulos TN et al (2004) Multiple molecular determinants in the carboxyl terminus regulate dopamine transporter export from endoplasmic reticulum. J Biol Chem 279:30760–30770
32. Farhan H, Korkhov VM, Paulitschke V et al (2004) Two discontinuous segments in the carboxy terminus are required for membrane targeting of the rat GABA transporter-1 (GAT1). J Biol Chem 276:28553–28563
33. Farhan H, Reiterer V, Korkhov VM et al (2007) Concentrative export from the endoplasmic reticulum of the gamma-aminobutyric acid transporter 1 requires binding to SEC24D. J Biol Chem 282:7679–7689
34. Farhan H, Reiterer V, Kriz A et al (2008) Signal-dependent export of GABA transporter 1 from the ER-Golgi intermediate compartment is specified by a C-terminal motif. J Cell Sci 121:753–761
35. Reiterer V, Maier S, Sitte HH et al (2008) Sec24- and ARFGAP1-dependent trafficking of GABA transporter-1 is a prerequisite for correct axonal targeting. J Neurosci 28:12453–12464
36. El-Kasaby A, Just H, Malle E et al (2010) Mutations in the carboxyl-terminal SEC24 binding motif of the serotonin transporter impair folding of the transporter. J Biol Chem 285:39201–39210
37. Sucic S, El-Kasaby A, Kudlacek O et al (2011) The serotonin transporter is an exclusive client of the coat protein complex II (COPII) component SEC24C. J Biol Chem 286:16482–16490

38. Sucic S, Koban F, El-Kasaby A et al (2013) Switching the clientele: a lysine residing in the C terminus of the serotonin transporter specifies its preference for the coat protein complex II component SEC24C. J Biol Chem 288:5330–5341
39. Torres GE, Yao WD, Mohn AR et al (2001) Functional interaction between monoamine plasma membrane transporters and the synaptic PDZ domain-containing protein PICK1. Neuron 30:121–134
40. Bjerggaard C, Fog JU, Hastrup H et al (2004) Surface targeting of the dopamine transporter involves discrete epitopes in the distal C terminus but does not require canonical PDZ domain interactions. J Neurosci 24:7024–7036
41. Madsen KL, Thorsen TS, Rahbek-Clemmensen T et al (2012) Protein interacting with C kinase 1 (PICK1) reduces reinsertion rates of interaction partners sorted to Rab11-dependent slow recycling pathway. J Biol Chem 287:12293–12308
42. Carneiro AM, Ingram SL, Beaulieu J-M et al (2002) The multiple LIM domain-containing adaptor protein Hic-5 synaptically colocalizes and interacts with the dopamine transporter. J Neurosci 22:7045–7054
43. Fog JU, Khoshbouei H, Holy M et al (2006) Calmodulin kinase II interacts with the dopamine transporter C terminus to regulate amphetamine-induced reverse transport. Neuron 51:417–429
44. Navaroli DM, Stevens ZH, Uzelac Z et al (2011) The plasma membrane-associated GTPase Rin interacts with the dopamine transporter and is required for protein kinase C-regulated dopamine transporter trafficking. J Neurosci 31:13758–13770
45. Wei Y, Williams JM, Dipace C et al (2007) Dopamine transporter activity mediates amphetamine-induced inhibition of Akt through a Ca2+/calmodulin-dependent kinase II-dependent mechanism. Mol Pharmacol 71:835–842
46. Speed NK, Matthies HJ, Kennedy JP et al (2010) Akt-dependent and isoform-specific regulation of dopamine transporter cell surface expression. ACS Chem Neurosci 1:476–481
47. Stöckli J, Fazakerley DJ, James DE (2011) GLUT4 exocytosis. J Cell Sci 124: 4147–4159
48. St Pierre CA, Leonard D, Corvera S et al (2011) Antibodies to cell surface proteins redirect intracellular trafficking pathways. Exp Mol Pathol 91:723–732
49. Tampellini D, Magrane J, Takahashi RH et al (2007) Internalized antibodies to the Abeta domain of APP reduce neuronal Abeta and protect against synaptic alterations. J Biol Chem 282:18895–18906

Chapter 10

Organic Cation Transport Measurements Using Fluorescence Techniques

Giuliano Ciarimboli and Eberhard Schlatter

Abstract

Analysis of transport processes using fluorescent substrates is a powerful tool to dynamically study various aspects of several transporters. The fluorescent organic cation 4(4-dimethylaminostyryl)-*N*-methylpyridinium (ASP^+) due to its specific fluorescence properties is a valuable probe for studying transport mediated by transporters of organic cations. ASP^+ is accepted by many members of the family of organic cation transporters and can be utilized in a wide spectrum of experimental settings analyzing such transport dynamics from the in vitro to the in vivo situation. Since its first introduction in 1994, ASP^+ has been widely used by several laboratories worldwide for all kinds of studies analyzing organic cation transport.

Key words ASP^+, Organic cation transport, OCT, Proximal tubule, Microtiter plate reader, Epifluorescence, Fluorescence imaging, Two-photon microscopy, Expression systems

1 Introduction

Transport of organic anions and cations across cell membranes of epithelial or neuronal cells of various organs and tissues is mediated by specific transport systems. These transport processes were initially studied either by the use of radioactively labeled substrates, electrophysiological analysis of electrogenic substrate transport, or by use of biochemical substrate analyses such as chromatography or mass spectroscopy. All these methods have their limitations as the availability of radiolabeled substances is costly and limited, electrophysiological analysis is restricted to electrogenic transporters, and especially biochemical or isotopic analyses do not allow continuous and dynamic analyses of transport processes. A more recently introduced alternative is the dynamic analysis of transport of fluorescent organic substrates using various epifluorescence techniques. In this chapter we summarize and discuss the development and technical aspects of the fluorescence methods which have been developed to study organic cation transport in various biological systems in vivo and in vitro.

Heinz Bönisch and Harald H. Sitte (eds.), *Neurotransmitter Transporters: Investigative Methods*, Neuromethods, vol. 118,
DOI 10.1007/978-1-4939-3765-3_10,

The first reports on the use of a fluorescent substrate to record organic substrate transport appeared in the 1970s when fluorescein was used to analyze organic anion transport in the frog kidney [1, 2] or the rat kidney [3]. It took 20 more years before the first report on the use of a fluorescent organic cation to study organic cation transport was reported [4, 5]. In these studies besides sulfofluorescein as fluorescent organic anion the fluorescent organic cation 4(4-dimethylaminostyryl)-*N*-methylpyridinium (ASP^+) was introduced to record organic substrate transport in the rat kidney in situ. Since then epifluorimetric or two-photon microscopic analysis of organic cation transporters using ASP^+ has been further developed and extended to various cell systems, organs, and species. These studies involve characterization of for example substrate affinities, structure–function relations, involvement in drug transport and toxicity, transport regulation by protein kinases and molecular interaction of transport proteins with other cytosolic proteins (for review see refs. [6–9]). Major advantages of this method are the commercially availability of ASP^+, the specific fluorescent properties of ASP^+, the acceptance of ASP^+ as high affinity substrate by most organic cation transport systems, the possibility of dynamic and real time transport analysis, or the relatively simple detection techniques (e.g., epifluorescence microscopy, microtiter plate reader analysis). These aspects and the various application modes will be discussed in detail in the next sections.

2 Properties of ASP^+ as Fluorescent Substrate

ASP^+ is a fluorescent probe, which belongs to the class of styryl dyes and can be purchased from Molecular Probes/Invitrogen (Fig. 1). These molecules are characterized by the presence of both electron donating and electron accepting groups, which are coupled through a motif of alternating single and double bonds [10, 11]. Due to this configuration, styryl dyes display in the excited electronic states a strong intramolecular charge transfer character. This causes prominent solvent polarity and viscosity dependent

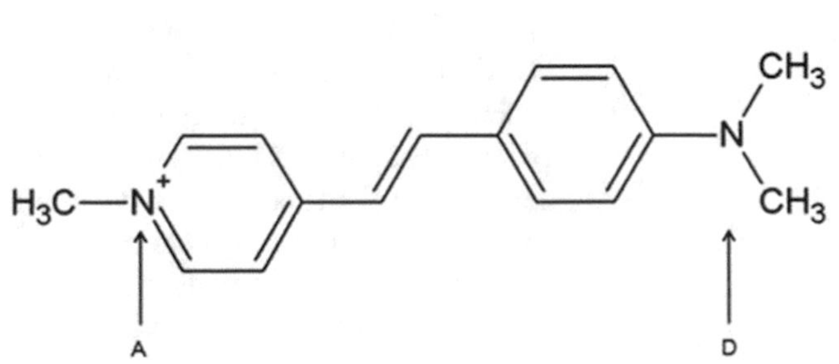

Fig. 1 Chemical structure of 4(4-dimethylaminostyryl)-*N*-methylpyridinium (ASP^+). The electron donor (D) and acceptor (A) groups are indicated

changes in their emission characteristics, undergoing large Stokes' shift in the fluorescent spectra and significant changes in the fluorescence quantum yields and lifetimes when the solvent polarity changes [10]. The polarity-based influences manifest themselves as a red shift in the emission spectra and a decrease in quantum efficiency and lifetime in the case of increasing polarity of the environment. An increase in viscosity results in reduced torsional dynamics, increased lifetimes, and enhancement of fluorescence at the red edge of the emission spectrum [12]. Specifically, ASP^+ is weakly fluorescent in aqueous solution, but its fluorescent quantum yield increases significantly in viscous media and upon adsorption on biological membranes [13], which also causes an emission spectrum shift [14].

Because of its fluorescent properties, ASP^+ is widely used for the investigation of transmembrane potentials of living cells [13], or for mitochondria staining [15]. Moreover, since ASP^+ has a p*Ka* of 3.8 [5], it is positively charged at pH 7.4, it has a low lipophilicity, and is not toxic. For these reasons it is an ideal substrate for the study of the activity of organic cation transporters [4, 5].

3 Fluorescence Analysis of Organic Cation and Anion Transport In Vivo

Already in 1976 Steinhausen et al. [3] introduced sulfofluorescein as a fluorescent organic anion suitable to study renal organic anion transport. These authors applied intravital fluorescence microscopy and micro photometry to the surface of the rat kidney in situ to quantify tubular secretion of organic anions. In 1994, Rohlicek and Ullrich described a simple device for continuous measurements of organic cation transport in the rat kidney in situ using the aminostyryl compound ASP^+ [4]. The protein binding of ASP^+ was lower compared to the fluorescent substrate rhodamine 123 in plasma (47% versus 72%), which was also tested by these authors. They used a halogen lamp together with various optical filters, a heat absorption filter, and a condenser as illumination unit to guide the specific excitation light (470 nm) via fiber optics to the surface of a rat kidney in situ. The light emitted (495 nm) from the kidney surface was detected after passing through a band pass filter by a photodiode as detector and recorded on a chart recorder. Together with classical peritubular and luminal perfusion experiments with specific substrates for luminal and basolateral organic cation transporters and fluorimetric analysis of the fluorescent substrates in the urine the authors demonstrated that substrates of the luminal and basolateral organic cation transporters interfere with ASP^+ transport and that ASP^+ is a substrate for both luminal and basolateral organic cation transporters in the rat proximal tubule.

Later the fluorimetric analysis of organic substrate transport was improved by using two-photon fluorescence microscopy which allows collecting fluorescence emissions within a small optical section. This technique was applied both to isolated proximal tubules [16] as well as to the rat kidney in vivo [17, 18]. In the latter studies glomerular filtration and tubular reabsorption along the nephron in the rat in vivo was quantified for the organic anion sulfofluorescein [17] or the organic cation ASP^+ [16, 18]. They demonstrated in their studies using freshly isolated rat proximal tubules, that ASP^+ is concentration dependently (1–50 μM) taken up across the basolateral membrane and that its transport can be inhibited by known competitive inhibitors of organic cation transport (N1-methylnicotinamide, tetraethyl ammonium, or choline). They suggested that this method is a useful tool for investigation of drug transport and interactions allowing continuous measurement of organic cation transport also under pathophysiological conditions.

Besides these few fluorimetric in vivo studies on organic substrate transport in the rat kidney, the interest of the researchers in this field now focused more on molecular and functional work on organic substrate transporters in expressions systems such as the *Xenopus* oocyte or HEK293 cells. Only relatively few studies have been performed ex vivo in freshly isolated renal proximal tubules or hepatocytes.

4 Organic Cation Transport in Proximal Tubule Cell Lines

Using the fluorescent organic cation ASP^+ we adapted the approach by Rohlicek and Ullrich [4] and characterized organic cation transport across the luminal membrane of pork and human proximal tubule cells lines [19–23]. We were able to monitor dynamically the concentration dependent uptake of ASP^+ across the apical membrane of these cell lines. Competitive inhibition of ASP^+ uptake by these cells was demonstrated for cimetidine, choline, or various quaternary ammonium compounds as well as its dependence on the membrane potential. In first experiments with these cells we demonstrated interaction of cytostatics and neurotransmitters with ASP^+ uptake across the apical membrane. In these studies we further examined for the first time the acute regulation of these transporters by various protein kinases. The cells were grown on glass coverslips to confluency. These coverslips were mounted as the bottom of a perfusion chamber in an inverted microscope (Axiovert, Zeiss, Oberkochen, Germany) and constantly superfused with experimental Ringer like solutions (e.g., in mM: NaCl 145, K_2HPO_4 1.6, KH_2PO_4 0.4, d-glucose 5, $MgCl_2$ 1, calcium gluconate 1.3, and pH adjusted to 7.4). The excitation light was generated by a xenon lamp (XBO 75 W, Zeiss) and after passing through a 450–490 nm band pass filter mounted on a filter wheel to generate pulsating light it was directed onto the cells via a dichroic filter (560 nm). The emitted light was collected by a photomultiplier after passing through a 575–640 nm band pass filter [21]. The control of the experiments and their analysis was reached using a computer-aided system and specific software (U. Fröbe, University of Freiburg, Germany). As a measurement of ASP^+ uptake, the increase of cellular fluorescence is plotted versus time. The signals of ten pulses every second are averaged, yielding a time resolution of 1 Hz. Because ASP^+ fluorescence is bleached by light, the whole device has to be protected from light. ASP^+ uptake can be recorded for several minutes, allowing the estimation of the maximal cellular fluorescence reached as well as of the initial slope of the fluorescence increase (linearly fitted during the first 30 s) as transport parameters. Background fluorescence measured for each monolayer in the absence of ASP^+ was subtracted from every experiment. The number of cells under analysis can be varied using an adjustable iris diaphragm. With this setup we demonstrated temperature and concentration dependent accumulation of ASP^+ across the apical membrane of these proximal tubular cells, which could be inhibited by another organic cation known to inhibit organic cation transporters. We also observed marked accumulation of ASP^+ by the mitochondria when cells were incubated with high concentrations of ASP^+ for at least 1 min. Therefore, to analyze ASP^+ transport mediated by plasma membrane transporters, only the initial rate of fluorescence increase with micromolar substrate concentrations was analyzed. Otherwise the fluorescence signal obtained after incubation of cells for several minutes is the sum of plasma membrane uptake, intracellular compartmentation, plasma membrane export, and fluorescence bleaching.

These experiments, although using cells which endogenously express organic cation transporters, have the disadvantage that different transporters are expressed in both the luminal and basolateral membrane and transport characteristics cannot easily be attributed to the activity of a distinct transporter. In the basolateral membrane of the native proximal tubular cells only the electrogenic organic cation transporter OCT is present. Which of the three subtypes of this transporter (OCT 1–3) is expressed varies between the species. While in rodents OCT1 and to a lesser extent OCT2 are expressed [24–27], in humans only OCT2 is present in this nephron segment [7, 28–30]. OCT3 seems to be expressed in the proximal tubule at very low levels in all species studied so far. In the luminal membrane various different organic cation transporters are expressed in this nephron segment [7, 8, 25]. Furthermore, we noticed after studying various available proximal tubular cell lines, that at least the expression of the organic cation transporters of the basolateral membrane (OCT 1–3) is apparently lost in these permanent cell lines. Therefore, for transport studies on OCTs it appears necessary to either use primary cell cultures of proximal tubules or expression systems which are transfected with OCTs (so far *Xenopus laevis* oocytes, HEK293, CHO, or MDCK cells have been used for this purpose). The methods to study these systems are described in the next sections.

Fig. 2 Confocal images of HEK293 cells stably expressing hOCT2 (*left*) and wild type cells (*right*) after incubation with 1 μM ASP^+

5 Organic Cation Transport Analyzed with a Photon Counting Tube or an Imaging System

Fluorescence microscopy experiments of ASP^+ uptake in cell lines used for the expression of distinct organic cation transporters were also performed with the setup described in Chap. 4 using an inverted microscope and a photon counting tube [19–22, 31]. Figure 2 depicts HEK293 cells which were stably transfected with hOCT2 and of wild type HEK293 cells after short incubation with ASP^+.

With this setup different subtypes of organic cation transporters from rat and man were characterized with respect to their apparent substrate affinities. In addition, for the first time acute regulation of these organic cation transporters by various protein kinase pathways were studied (for review see refs. [6, 7, 32]). Figure 3 shows original fluorescence recordings obtained with this setup for the effect of PKA activation by forskolin of rOCT1 stably expressed in HEK293 cells. Using this system we were also able to demonstrate direct phosphorylation of rOCT1 by PKC activation leading to changes in the apparent substrate affinity for ASP^+ [31, 33].

6 Organic Cation Transport Analyzed with a Microtiter Plate Fluorescence Reader

As described above, when ASP^+ is taken up by cells it displays a shift in its emission spectrum (Fig. 4). Taking advantage of this phenomenon, we developed a microfluorimetrical detection system using a fluorescence plate reader (Infinity M200, Tecan, Crailsheim, Germany). Such a system allows performing high throughput dynamic measurements of ASP^+-uptake. Other known fluorescent substrates for OCTs that do not show such an emission spectrum shift, as for example amiloride, are no suitable substrates

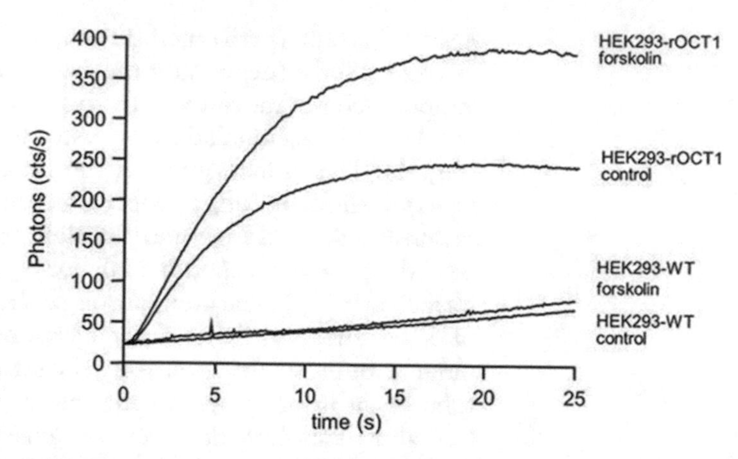

Fig. 3 Original recordings of ASP+ (1 μM) fluorescence increase in wild type (WT) or rOCT1 stably transfected HEK293 cells. Cells were incubated with 1 μM forskolin for 10 min. Data are means of several identical experiments

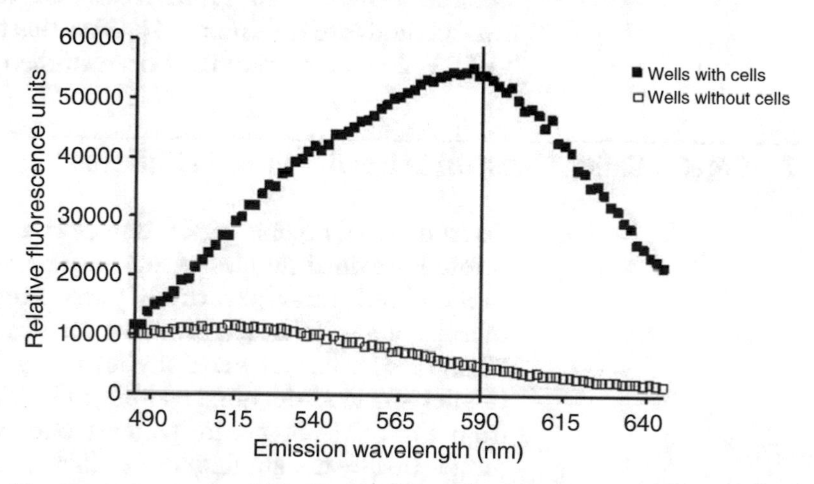

Fig. 4 Emission spectrum of ASP+ (1 μM) at an excitation wavelength of 450 nm in wells containing (*closed squares*) or not (*open squares*) HEK-293 cells stably transfected with rOCT2. The *vertical line* indicates the wavelength chosen for fluorescence emission readings. Modified from ref. [9]

for such a high throughput measurement system [14]. Briefly, to perform these fluorescence measurements, cells confluently grown on 96-well-microplates (Nunclon 96 Flat bottom, Nunc, Wiesbaden, Germany) have to be placed in uptake buffer tempered at 37 ° C. The microtiter plate with the cells is then inserted into the fluorescence reader. To optimize the time resolution maximally four wells are excited with monochromatic light of 450 nm over a time period of up to 180 s. Fluorescence emission, filtered by a second monochromator at 590 nm, is finally measured by a fluorescence detector. Fluorescence is measured in each well for three consecutive data points to obtain a stable background signal before

ASP^+ is injected (performed by an injector) with a time resolution of around 0.6 s (depending on the number of wells that are contemporaneously measured). In this way a dynamic registration of cellular ASP^+-accumulation as result of ASP^+-uptake is obtained (Fig. 4). The sudden jump in the fluorescence signal after ASP^+ injection allows to judge proper function of the injector. The fluorescence detection time and the time resolution can be changed, according to the experimental needs. All experiments are performed at 37 °C. The comparison of the IC_{50} values for inhibition of ASP^+ uptake by other organic cations obtained with this technique with those obtained using the setup with a photon counting tube or an imaging system mounted on an inverted microscope (see above) showed that the two analysis systems give the same qualitative and also quantitative results [14]. When investigating regulation pathways in the fluorescence reader system, an incubation step with the respective agonists or inhibitors has to be added before injection of ASP^+. Also in this case, results obtained with the two methods are the same [14]. With this method meanwhile also OCT 1, 2 and 3 of mice have been studied extensively [34, 35].

7 Organic Cation Transport in Isolated Proximal Tubules

To study renal organic cation transport as it occurs in vivo freshly isolated proximal tubules of rat, mice or humans have been used. In these studies transport of ASP^+ across the basolateral membrane of isolated and collapsed proximal tubules has been analyzed [16, 27, 36]. The first of these studies was performed with proximal tubules isolated from human kidneys [28]. This tissue was obtained from tumor nephrectomy patients and was comprised of non-tumor tissue surrounding the kidney tumor. Proximal tubules, mostly straight S3 segments, were mechanically isolated from thin slices which were cut from the tissue samples using fine forceps. Single tubule segments were transferred to a micro perfusion chamber which was mounted on the stage of an inverted microscope (Axiovert, Zeiss) equipped with the fluorescence analysis setup as described in Sect. 4. The collapsed tubule segments were fixed by aspiration of both ends by two fine glass pipettes mounted on micromanipulators (see . 5).

This approach immobilizes the tubule segment in the constantly perfused and temperature controlled chamber and also prevents an interference of ASP^+ or any of the applied drugs or factors with the luminal membrane of the tubules which expresses other organic cation transporters. As the human proximal tubule expresses only the OCT subtype 2 (hOCT2) in the basolateral membrane, the recorded uptake of ASP^+ from the bath solution can be entirely attributed to this transporter. This is different compared to the situation in isolated proximal tubules of rats and mice, which express two subtypes each (mOCT1 and mOCT2 or rOCT1 and rOCT2) in this membrane domain. In the human proximal tubules hOCT2 mediated transport was characterized as membrane voltage dependent and other organic substrates or inhibitors of OCTs interfered with ASP^+ uptake in these tubules with K_i values comparable to those calculated for hOCT2 expressed in HEK293 cells (see Chaps. 5 and 6). In these studies also the acute regulation of hOCT2-mediated ASP^+ uptake by various protein kinases was analyzed (Fig. 6). These results correspond well with those obtained for hOCT2 expressed in HEK293 cells and differ markedly from those obtained for hOCT1, the subtype expressed in the liver but not in the kidney. We have also adapted this system for the study of ASP^+ uptake in freshly isolated human hepatocytes [37]. In this case, hepatocyte couplets forming a canalicular space and in this way retaining their polarity were placed in the perfusion chamber and held by aspiration in a fine glass pipette. The detection window was placed on the sinusoidal membrane of hepatocytes, where hOCT1 is expressed.

Similarly characteristics of organic cation transport mediated by OCTs in the basolateral membrane of freshly isolated proximal tubules of mouse or man have also been performed by us using the

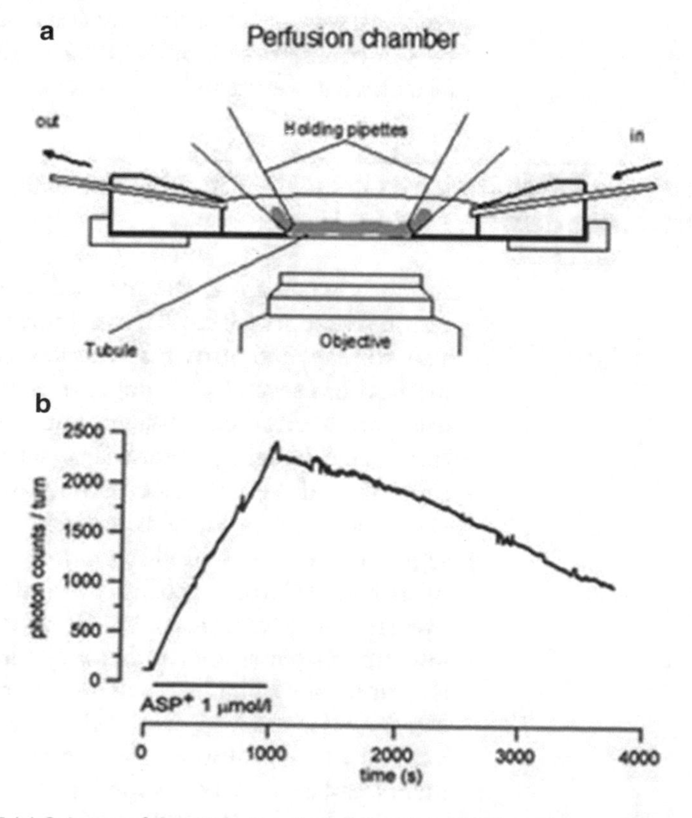

Fig. 5 (**a**) Scheme of the perfusion chamber with isolated proximal tubule and (**b**) original recording of cellular ASP^+ fluorescence after uptake across the basolateral membrane of an isolated human proximal tubule

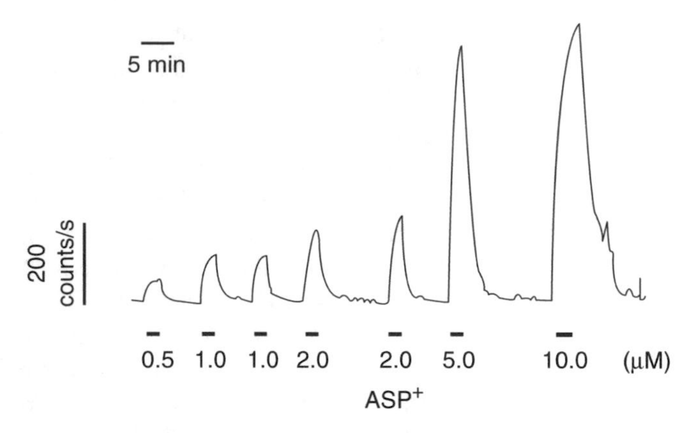

Fig. 6 Concentration dependent cellular ASP+ fluorescence increase of an isolated human proximal tubule

same experimental setup, but instead of a photon counting tube an imaging system (Visitron, Puchheim, Germany) with a CCD camera was used [27, 38]. In this system data acquisition and analysis were done with Metafluor Software (Visitron). The excitation light (488 nm) was generated by a polychromator system (VisiChrome) and reflected to the preparation by a dichroic mirror (560 nm) and emission was detected after passing a band pass filter (575–640 nm) by a Photometrics CoolSNAP EZ digital camera (Roper Scientific, Martinsried, Germany).

8 Organic Cation Transport in Isolated Tubules Analyzed with a Microtiter Plate Fluorescence Reader

Recently we adapted the method to dynamically analyze organic cation transport with ASP+ in a fluorescence microtiter plate reader also for isolated proximal tubules of mice [27, 35, 36]. This method has several advantages over the above described methods using an inverted fluorescence microscope and a photon counting tube or an imaging system. The setup procedure for each single tubule is tedious and time-consuming in the inverted microscope system and needs some experience in handling such small tubule segments (approx 100–300 μm long, 10–15 μm in diameter) with micromanipulators. Setting up and performing a single tubule experiment takes at least 15–30 min each, allowing only to use a few tubules per animal and kidney slices have to be kept on ice for isolation for several hours of an experimental day. For the fluorescence reader method many tubule segments are isolated from the kidneys of each mouse (e.g., up to 800 S3-proximal tubule segments isolated by two experienced experimentators within 2 h) resulting in a large number of experimental data for each animal. This allows reducing the number of animals necessary for such a

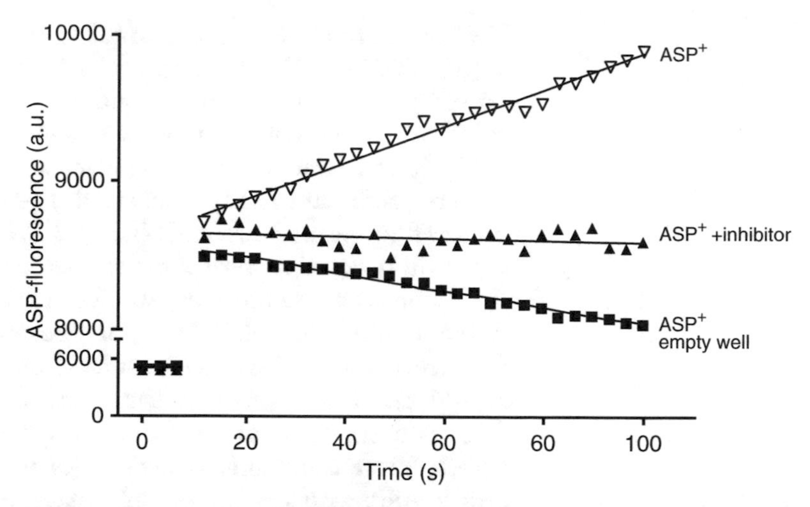

Fig. 7 Original recording of ASP+ (1 μM) fluorescence of wells containing three tubules each with or without transport inhibitor (TPA+, 1 mM) or an empty well of a 384-well microtiter plate measured in a Tecan fluorescence reader

study which is especially important if expensive and in availability limited genetically manipulated animals are used. The handling of the tubules is technically easy and the transport measurements can be performed in a semi-automatic way. After mechanical isolation of the tubules three segments (each set having a comparable total length) are transferred in a well of a 384-well microtiter plate (Deep Well Small Volume, Greiner, Frickenhausen, Germany) with 10 μl of experimental solution. Plates were mildly centrifuged to ensure settling of tubule segments at the flat bottom of the wells and to remove air bubbles. If needed, for example, for regulation experiments agonists or antagonists were added to the respective wells and incubation time was set as needed before the ASP+ uptake measurements were started. As described in Chap. 7 ASP+ is injected after sampling three consecutive baseline data points and fluorescence increase is analyzed during the initial linear phase. To correct for background fluorescence changes in this signal with time (bleaching) are recorded in empty wells and subtracted from the signals obtained in wells containing tubules (Fig. 7).

9 ASP+ as Fluorescent Marker for the Study of Transporters Other than OCTs

Transporters for organic cations are polyspecific transporter proteins, which share several substrates. For this reason, ASP+ could be a potential fluorescent marker of dynamic transport by such transporters. Indeed, ASP+ has been demonstrated to be a substrate of the human novel organic cation transporter type 1 (hOCTN1)

[39] and of the human multidrug and toxin extrusion transporter 1 (hMATE1) [39]. Interestingly, OCTN2, a transporter closely related to OCTN1, did not recognize ASP^+ as substrate [40]. Since ASP^+ is structurally similar to 1-methyl-4-phenylpyridinium (MPP^+), a substrate of norepinephrine and also of dopamine transporters (NET and DAT, respectively) [41, 42], it has been also used as fluorescent marker of NET and DAT activity. In the following, a short description of these studies is presented.

Fluorescence microscopy studies in HEK293 cells stably transfected with human NET, DAT, or serotonin transporters (SERT) performed using a Zeiss 410 confocal microscope with excitation at 488 nm and emission detected at 580–630 nm showed that ASP^+ is a substrate of all these transporters [43]. Focusing on hNET, fluorescence anisotropy studies performed by exposing the cells to ASP^+ with a polarizer rapidly switching from the horizontal to the vertical position and measuring light intensity in the horizontal and vertical positions, the authors detected two ASP^+ populations with distinct rotational movements: a rapid phase I, which is relative immobile and spatially localized to the cell surface and a slow phase II, which may represent mobile ASP^+ [43]. Analysis of spatial patterning and temperature dependence together with pharmacological dissection of the two phases suggests that phase I measures ASP^+ binding to plasma membranes and phase II measures ASP^+ transport. Moreover, by fluorescence studies in the presence of hydrophilic and hydrophobic quenchers, the authors showed that bound ASP^+ is unavailable to these quenchers, suggesting that it segregates from bath and membrane and that it binds deep within the transporter [43]. Experiments with superior cervical ganglia cells demonstrated that ASP^+ can be also used in cells naturally expressing the NET to monitor the activity of the transporter [43]. Since these cells express also OCT, the authors suggest that the part of ASP^+ uptake mediated by NET can be evaluated in the presence of 10 μM desipramine, which at this concentration should not significantly interact with OCT.

ASP^+ has been shown to be a suitable substrate for measuring the substrate–transporter stoichiometry because it allows separation of plasma membrane-bound from transported substrate [44]. In fact, using fluorescence lifetime microscopy (FLIM) and total intern reflection fluorescence (TIRF) a stoichiometry of 1:1 between ASP^+ and GFP-tagged hNET was determined [44]. Fluorescence correlation spectroscopy experiments combined with mass spectrometry revealed that a single ASP^+ molecule binds and unbinds thousands of times before being transported or ultimately dissociated from hNET, with a transport rate of 0.06 ASP^+/hNET-protein/s.

A similar approach was used to study regulation of the DAT function [45]. Here, cells stably expressing a tagged DAT were incubated with 10 μM ASP^+, and the fluorescence was captured using excitation at 488 nm and emission at 607–652 nm. The rate of ASP^+ uptake was calculated by determining the slope of the linear accumulation function for 1 min. The ASP^+ uptake by DAT resulted to be linear over the first 10 min after ASP^+ addition and to be temperature-dependent. For regulation studies, cells were incubated 15 min at 37 ° C with different kinase inhibitors before ASP^+ addition.

Recently, a detailed analysis of ASP^+ transport by hSERT transfected in EM4 cells, a variant of HEK293 cells stably expressing a macrophage scavenger to increase adherence to cell culture plastic, has been published [46]. Here time-resolved live cell images were obtained using a PerkinElmer LCI confocal system on a Nikon Eclipse 2000-S inverted microscope equipped with PlanApo objectives (PerkinElmer, Waltham, MA, USA). Images were collected at room temperature using a 60× objective and captured using a Hamamatsu CCD camera (Hamamatsu, Bridgewater, NJ, USA). Cells were rinsed with uptake buffer before addition of 10 μM ASP^+, which was excited by an argon laser at 488 nm. Fluorescence was captured at 607–652 nm. The system collected images every 200 ms. Nonspecific fluorescence was evaluated in non-transfected cells in the same dishes. To correct for background, initial ASP^+ fluorescence determined before ASP^+ addition was determined. Upon exposure of the cells to ASP^+, an immediate and persistent increase in fluorescence intensity was seen, also here characterized by an initial phase of short duration (v2.5 s), compatible with binding to the plasma membrane, followed by a second phase of fluorescence increase at a slower rate. The fluorescence observed in both phases correlated with the hSERT expression level. Also in this experimental setup, the regulation of hSERT by several different kinases was evaluated by treating the cells with activators or inhibitors for different time before addition of ASP^+, analogously to what is described in the regulation studies of OCT [31].

10 Conclusions

These studies confirm the special suitability of ASP^+ as fluorescent marker of cellular organic cation transport processes, allowing a temporal and spatial resolution of transport dynamics of a variety of organic cation transporters. Moreover, the wide spectrum of methodological applications, such as epifluorescence detected by photon counting tubes or digital cameras, two photon fluorescence microscopy, or fluorescence microtiter plate readers, allows to study organic cation transport dynamically in various in vivo, ex vivo, or in vitro settings. By using ASP^+ as substrate for these transporters further important biophysical information, such as transport stoichiometry and kinetics, can be obtained.

Acknowledgements

The work from the authors' laboratory described here was supported by the German Research Council (Schl 277/8-1 to 8-4 and 12-3 and CI 107/4-1 to 4-3), the German Krebshilfe Foundation (#108539), the Interdisciplinary Center of Clinical Research (IZKF, Cia2/013/13), and the Innovative Medical Research (IMF) of the Medical Faculty of the University Münster, Germany.

References

1. Bresler VM, Bresler SE, Nikiforov AA (1975) Structure and active transport in the plasma membrane of the tubules of frog kidney. Biochim Biophys Acta 406:526–537
2. Bresler VM, Natochin I (1973) Diuretic inhibition of fluorescein secretion in the proximal kidney tubules of the frog (a study during life by the contact microscopy method). Biull Eksp Biol Med 75:67–69
3. Steinhausen M, Müller P, Parekh N (1976) Renal test dyes IV. Intravital fluorescence microscopy and microphotometry of the tubularly secreted dye sulfonefluorescein. Pflugers Arch 364:83–89
4. Rohlicek V, Ullrich KJ (1994) Simple device for continuous measurement of fluorescent anions and cations in the rat kidney in situ. Ren Physiol Biochem 17:57–61
5. Pietruck F, Ullrich KJ (1995) Transport interactions of different organic cations during their excretion by the intact rat kidney. Kidney Int 47:1647–1657
6. Ciarimboli G, Schlatter E (2005) Regulation of organic cation transport. Pflugers Arch 449:423–441
7. Ciarimboli G (2008) Organic cation transporters. Xenobiotica 38:936–971
8. Koepsell H, Endou H (2004) The SLC22 drug transporter family. Pflugers Arch 447:666–676
9. Koepsell H (2004) Polyspecific organic cation transporters: their functions and interactions with drugs. Trends Pharmacol Sci 25:375–381
10. Shaikh M, Mohanty J, Singh PK et al (2010) Contrasting solvent polarity effect on the photophysical properties of two newly synthesized aminostyryl dyes in the lower and in the higher solvent polarity regions. J Phys Chem A 114:4507–4519
11. Haidekker MA, Brady TP, Lichlyter D et al (2005) Effects of solvent polarity and solvent viscosity on the fluorescent properties of molecular rotors and related probes. Bioorg Chem 33:415–425
12. Ramadass R, Bereiter-Hahn J (2007) Photophysical properties of DASPMI as revealed by spectrally resolved fluorescence decays. J Phys Chem B 111:7681–7690
13. Glazachev YI, Semenova AD, Kryukova NA et al (2012) Express method for determination of low value of trans-membrane potential of living cells with fluorescence probe: application on haemocytes at immune responses. J Fluoresc 22:1223–1229
14. Wilde S, Schlatter E, Koepsell H et al (2009) Calmodulin-associated post-translational regulation of rat organic cation transporter 2 in the kidney is gender dependent. Cell Mol Life Sci 66:1729–1740
15. Villa AM, Doglia SM (2004) Mitochondria in tumor cells studied by laser scanning confocal microscopy. J Biomed Opt 9:385–394
16. Pietruck F, Hörbelt M, Feldkamp T et al (2006) Digital fluorescence imaging of organic cation transport in freshly isolated rat proximal tubules. Drug Metab Dispos 34:339–342
17. Tanner GA, Sandoval RM, Dunn KW (2004) Two-photon in vivo microscopy of sulfonefluorescein secretion in normal and cystic rat kidneys. Am J Physiol Renal Physiol 286:F152–F160
18. Hörbelt M, Wotzlaw C, Sutton TA et al (2007) Organic cation transport in the rat kidney in vivo visualized by time-resolved two-photon microscopy. Kidney Int 72:422–429
19. Hohage H, Stachon A, Feidt C et al (1998) Regulation of organic cation transport in IHKE-1 and LLC-PK_1 cells. Fluorimetric studies with 4-(4-dimethylaminostyryl)-N-methylpyridinium. J Pharmacol Exp Ther 286:305–310
20. Hohage H, Stachon A, Feidt C et al (1998) Effects of protein kinase activation on organic cation transport in human proximal tubular cells. Nova Acta Leopoldina NF 306:293–298
21. Stachon A, Schlatter E, Hohage H (1996) Dynamic monitoring of organic cation transport processes by fluorescence measurements in LLC-PK_1 cells. Cell Physiol Biochem 6:72–81

22. Stachon A, Hohage H, Feidt C et al (1997) Characterization of organic cation transport across the apical membrane of proximal tubular cells with the fluorescent dye 4-Di-1-ASP. Cell Physiol Biochem 7:264–274
23. Stachon A, Hohage H, Feidt C et al (1998) Cytostatics and neurotransmitters are transported by the organic cation transporter in proximal cells. Nova Acta Leopoldina NF 78(306):333–338
24. Gorboulev V, Ulzheimer JC, Akhoundova A et al (1997) Cloning and characterization of two human polyspecific organic cation transporters. DNA Cell Biol 16:871–881
25. Koepsell H (1998) Organic cation transporters in intestine, kidney, liver, and brain. Annu Rev Physiol 60:243–266
26. Okuda M, Saito H, Urakami Y et al (1996) cDNA cloning and functional expression of a novel rat kidney organic cation transporter, OCT2. Biochem Biophys Res Commun 224:500–507
27. Holle SK, Ciarimboli G, Edemir B et al (2011) Properties and regulation of organic cation transport in freshly isolated mouse proximal tubules analyzed with a fluorescence reader-based method. Pflugers Arch 462:359–369
28. Pietig G, Mehrens T, Hirsch JR et al (2001) Properties and regulation of organic cation transport in freshly isolated human proximal tubules. J Biol Chem 276:33741–33746
29. Urakami Y, Akazawa M, Saito H et al (2002) cDNA cloning, functional characterization, and tissue distribution of an alternatively spliced variant of organic cation transporter hOCT2 predominantly expressed in the human kidney. J Am Soc Nephrol 13:1703–1710
30. Motohashi H, Sajurai Y, Saito H et al (2002) Gene expression levels and immunolocalization of organic ion transporters in the human kidney. J Am Soc Nephrol 13:866–874
31. Mehrens T, Lelleck S, Çetinkaya I et al (2000) The affinity of the organic cation transporter rOCT1 is increased by protein kinase C dependent phosphorylation. J Am Soc Nephrol 11:1216–1224
32. Ciarimboli G, Struwe K, Arndt P et al (2004) Regulation of the human organic cation transporter hOCT1. J Cell Physiol 201:420–428
33. Ciarimboli G, Koepsell H, Iordanova M et al (2005) Individual PKC-phosphorylation sites in organic cation transporter 1 determine substrate selectivity and transport regulation. J Am Soc Nephrol 16:1562–1570
34. Guckel D, Ciarimboli G, Pavenstädt H et al (2012) Regulation of organic cation transport in isolated mouse proximal tubules involves complex changes in protein trafficking and substrate affinity. Cell Physiol Biochem 30:269–281
35. Massmann V, Edemir B, Schlatter E et al (2014) The organic cation transporter 3 (OCT3) as molecular target of psychotropic drugs: transport characteristics and acute regulation of cloned murine OCT3. Pflügers Arch 466(3):517–527
36. Schlatter E, Klassen P, Massmann Vet al (2914) Mouse organic cation transporter 1 determines properties and regulation of basolateral organic cation transport in renal proximal tubules. Pflügers Arch 466(8):1581–1589
37. Ciarimboli G, Ludwig T, Lang D et al (2005) Cisplatin nephrotoxicity is critically mediated via the human organic cation transporter 2. Am J Pathol 167:1477–1484
38. Ciarimboli G, Lancaster CS, Schlatter E et al (2012) Proximal tubular secretion of creatinine by organic cation transporter OCT2 in cancer patients. Clin Cancer Res 18:1101–1108
39. Schmidt-Lauber C, Harrach S, Pap T et al (2012) Transport mechanisms and their pathology-induced regulation govern tyrosine kinase inhibitor delivery in rheumatoid arthritis. PLoS One 7:e52247
40. Grigat S, Fork C, Bach M et al (2009) The carnitine transporter SLC22A5 is not a general drug transporter, but it efficiently translocates mildronate. Drug Metab Dispos 37:330–337
41. Russ H, Gliese M, Sonna J et al (1992) The extraneuronal transport mechanism for noradrenaline (uptake2) avidly transports 1-methyl-4-phenylpyridinium (MPP^+). Naunyn Schmiedebergs Arch Pharmacol 346:158–165
42. Kitayama S, Shimada S, Uhl GR (1992) Parkinsonism-inducing neurotoxin MPP^+: uptake and toxicity in nonneuronal COS cells expressing dopamine transporter cDNA. Ann Neurol 32:109–111
43. Schwartz JW, Blakely RD, DeFelice LJ (2003) Binding and transport in norepinephrine transporters. Real-time, spatially resolved analysis in single cells using a fluorescent substrate. J Biol Chem 278:9768–9777
44. Schwartz JW, Novarino G, Piston DW et al (2005) Substrate binding stoichiometry and kinetics of the norepinephrine transporter. J Biol Chem 280:19177–19184
45. Bolan EA, Kivell B, Jaligam V et al (2007) D2 receptors regulate dopamine transporter function via an extracellular signal-regulated kinases 1 and 2-dependent and phosphoinositide 3 kinase-independent mechanism 1. Mol Pharmacol 71:1222–1232
46. Oz M, Libby T, Kivell B et al (2010) Real-time, spatially resolved analysis of serotonin transporter activity and regulation using the fluorescent substrate, ASP^+. J Neurochem 114:1019–1029

Chapter 11

Animal Models for Depression and the Mode of Action of Neurotransmitter Transporter-Blocking Antidepressants

Deeba Khan, Marianne Ronovsky, Thomas Steinkellner, Michael Freissmuth, Harald H. Sitte, and Daniela D. Pollak

Abstract

Major depressive disorder is a highly prevalent and devastating mental illness whose underlying pathomechanisms remain incompletely understood. Currently the most commonly used antidepressants block neurotransmitter transporters, i.e., the transporters for serotonin (SERT) and/or norepinephrine (NET); in addition, there are compounds which regulate transmitter release by targeting presynaptic autoreceptors for norepinephrine and serotonin. There is an unmet medical need, because some patients respond poorly to the existing treatments. Preclinical models are required to test potential new antidepressant compounds. These models must recapitulate at least some of the symptoms of depression in animals and allow for the quantification of treatment effects on depression-like behavior. Here we present the *chronic mild stress* model as a suitable paradigm for eliciting depression-like behavior in mice. We describe the procedures used to interrogate the model by quantifying the induced behavioral phenotype, the associated physiological alterations and their modulation by drug candidates.

Key words Animal model, Behavior, Depression, Neurotransmitter transporter, Anhedonia

1 Introduction and Background

Major depressive disorder (MDD; unipolar, clinical depression) is a highly prevalent and extremely devastating mental illness which additionally constitutes an important socioeconomic burden worldwide [1]. Currently most commonly used antidepressant pharmacological interventions are largely based upon the blockage of neurotransmitter transporters, i.e., the transporters for serotonin (SERT) and/or norepinephrine (NET). These drugs are classified as the very frequently prescribed specific serotonin reuptake inhibitors (SSRIs), specific norepinephrine reuptake inhibitors (SNRIs), and tricyclic antidepressants (TCA). Additional drug targets are available, in particular the presynaptic autoreceptors for norepinephrine and serotonin, which regulate neurotransmitter release. It is, however, obvious that there is an unmet medical need:

Heinz Bönisch and Harald H. Sitte (eds.), *Neurotransmitter Transporters: Investigative Methods*, Neuromethods, vol. 118, DOI 10.1007/978-1-4939-3765-3_11,

conservative estimates indicate that 30% of the patients do not achieve remission, even when offered up to four different sequential treatments [2]. The selectivity of the currently available drugs can be improved by approaches that are based on in vitro assays which reduces the number of side effects by minimizing off-target effects. However, in vitro assays cannot identify new candidate drug targets. This requires a better understanding of the pathophysiology underlying MDD. In addition, the systemic effects of new drugs can only be explored in an animal model, which is predictive of a response in human patients. Accordingly, the animal model ought to recapitulate some of the major features of MDD. Because of their importance for both, understanding the pathophysiology of depression and selecting drug candidates, animal models have been continuously refined over the past five decades.

Animals can be genetically modified to display a phenotype resembling depression; alternatively or in addition, they can be subjected to exogenous stimuli, such as stressful conditions, which induce a state mimicking the disorder. Lesions in specific parts of the brain and electrophysiological studies have offered clues as to the neuroanatomy and neural circuitry of depression. In addition to these, behavioral paradigms for the induction of depression-like phenotypes are also widely employed in mood disorder research (see for review: [3–5]). In order to mimic the human situation of uncontrollable life stressors implicated as precipitating factor for the development of MDD [6, 7], Willner and colleagues developed the chronic mild stress procedure (CMS) in the 1980s [8]. Herein, an animal is repetitively, over a designated period, exposed to ethically justifiable microstressors, such as olfactory exposure to a predator, despair stress, and psychological stress. The CMS model was originally studied in rats [9–12]; however, an increasing number of studies support the validity of this model in different mouse strains [2, 13–15].

The development of depression-like behavior is evaluated by recording consumption of palatable substances, such as sucrose solution, in order to reveal a potential impairment in reward processing, a decrease in sucrose consumption, suggesting an anhedonic phenotype. Anhedonia, defined as the loss of experiencing pleasure in normally pleasurable circumstances, is a cardinal feature in MDD patients and evaluated in laboratory animals using the sucrose preference test (SPT) [16, 17]. The present chapter describes a CMS paradigm as routinely employed in our laboratory [14] based upon the original description by Strekalova et al. (2004) [15]. The induced depression-like phenotype is assessed by the SPT and by evaluating fur state and changes in body weight [16].

This approach is not only of interest for drug development but also an indispensable tool for neurotransmitter transport research, when the goal is to link short and long term changes in expression levels of neurotransmitter transporters to behavioral outcomes.

2 Equipment, Materials, and Setup

2.1 Mouse Housing, Husbandry, and Handling

All experiments involving live animals need to be approved by the institutional and national ethical committee on animal care and use and should be carried out strictly adhering to international laws and policies with the aim to reduce pain and suffering and the number of animals used in each experiment. All behavioral procedures proposed have been validated in adult C57BL/6 N male mice (8–10 weeks of age, with an average body weight of 26–30 g at the beginning of the experiments).

Upon delivery by a certified vendor, mice are single housed in Macrolon Type II cages under a 12:12 h light–dark cycle (lights on: 7 am) using standard laboratory conditions (temperature 21 ± 1 °C, relative humidity 40–60 %) with access to food and water ad libitum, unless otherwise stated. Mice are allowed to habituate to the novel environment for a period of 2 weeks before the start of experiments.

2.2 Drugs of Choice

Various pharmacological agents can be used for the evaluation of their potential effects on depression-like behavior. Optimal dosage and treatment duration as well as the route of administration—oral administration (via food or drinking water) or injections [intraperitoneal (i.p.), intramuscular (i.m.), or subcutaneous (s.c.)]—are inferred from the literature or need to be determined before the actual experimental trials, considering the pharmacological properties of the respective substance (*Note 1*).

2.3 Equipment Parts and Assembly

(A) *CMS paradigm*

The CMS paradigm consists of exposure to three different kinds of stressors over a period of 28 days, which include olfactory exposure to a predator, psychological stress using spatial restraint and despair stress. For exposure to olfactory cues of a predator, experimental mice are individually housed in rectangular transparent plastic cages, size 21 × 16 × 12 cm. Eighteen holes (diameter 0.3 cm) are made on one long side of the box to allow for a diffusion of olfactory cues, in addition to 18 holes (0.3 cm diameter) on the lid of the cage. This ensures a secure enclosure for the mice when placed in the rat cage. Each box is then placed into a rat home cage (~5 mice boxes for one rat) of size of 116 × 56 × 51 cm, containing one Wistar rat, weighing on average 600 g (*Note 2*).

For spatial restraint stress, mice are placed inside a plastic transparent tube (internal diameter 2.6 cm). A 0.3 cm hole at the end of the tube and six additional holes along the length of the tube are made to support adequate delivery of air to the animal. The tubes are placed under a dark box of size 60 × 40 × 30 cm in order to create a dark environment, which eliminates additional stress from light exposure. Despair stress is evoked by suspending mice by their tails on a metal horizontal shaft by means of first aid waterproof tape for a period of 6 min.

(B) *Tests to evaluate depression-like phenotypes: SPT, fur state and body weight evaluation*

SPT

Drinking tubes are custom-made in order to accurately measure liquid consumption during the SPT. To this end, graduated 50 ml conical tubes (CellStar Tubes, Cat#:GRE-227-261, M&B Stricker, Tutzing, Germany) are cut on the conical side using a fine-tooth coping saw. A rubber stopper (Fisherbrand One Hold Rubber Stoppers, Size: 2, Hole diameter: 5 mm (Cat#: 1134-7291, Fisher Scientific Austria, Vienna, Austria)) is inserted into the conical tube and then a sipper tube (3″ Straight Stainless Steel Sipper Tube: (Cat#: 900007, Dyets Inc., Bethlehem, PA, USA)) is introduced into the rubber stopper. The rubber stoppers must be inserted into 50 ml conical tubes prior to placing the sipper tubes inside, otherwise they expand and no longer fit into the opening of the 50 ml tube (Fig. 1).

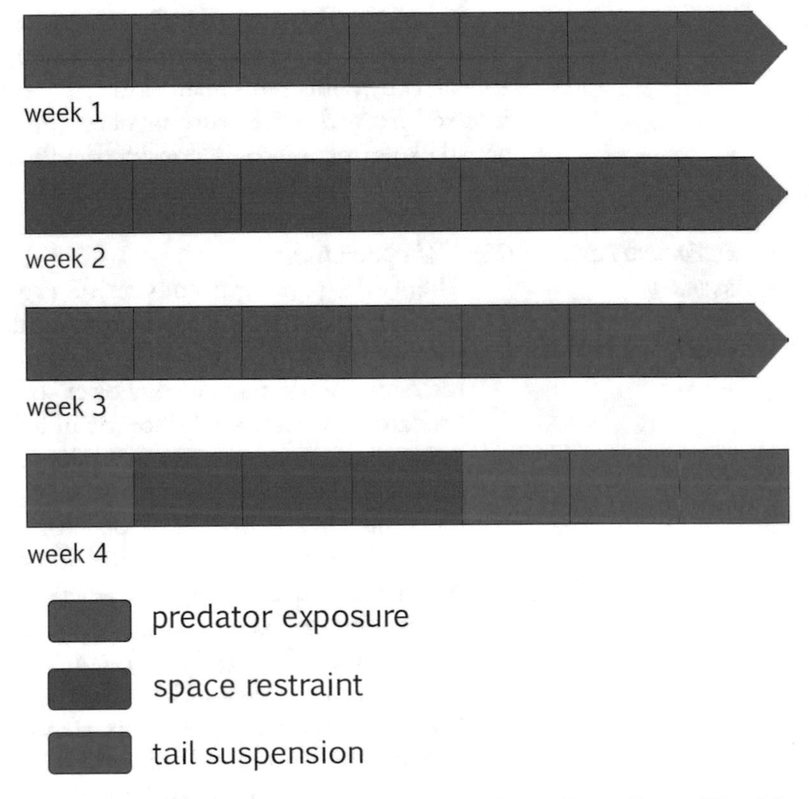

Fig. 1 Schematic representation of the experimental paradigm utilized for the chronic mild stress procedure. The 4 weeks (28 days) CMS paradigm consists of a daily exposure to one of three different stressor (predator exposure, space restraint and tail suspension) in varying sequences. The typical temporal course of stressor schedules is schematically depicted

A 2% sucrose (Cat# 16104-1KG, Sigma-Aldrich, St. Louis, MO) solution is prepared freshly and filled into one conical 50 ml tube. The other one is filled with normal drinking water. This allows for determining if the animals display a preference for sucrose solution compared to normal drinking water.

2.4 Fur State

The fur state is evaluated by direct observation of the experimenter. Scores are determined as previously described [18, 19] and listed below.

1. The fur is well-groomed. It appears shiny and smooth, without any disordered parts. Whiskers are of normal length and eyes have clear conjunctivae.
2. The fur is mostly like described above (see point 1), nevertheless some parts of the fur are disordered and fluffy.
3. The fur appears disordered and fluffy on most parts of the body and may have fine stains. Whiskers may be of aberrant length and eye conjunctivae may appear red.
4. The fur is disordered, fluffy, stained and soiled. Some bald patches or traces of wounds may exist. Some parts of the body may be hairless and wounds may be detected; eyes are red.

2.5 Body Weight Evaluation

Body weight is evaluated using an electronic scale (range 0–100 g, with a precision of 0.01 g) by putting the animal into a light plastic container, placed on the scale (*Note 3*).

3 Procedures

Seven days prior to initiating of the actual experiment, mice need to be familiarized to the experimenter by daily handling for a short period of time. Hereby, each animal is taken out of their cage by their tail and kept on the hand for around 2 min (*Note 4 and 5*).

3.1 Behavioral Protocols

(A) *CMS paradigm*

In the CMS paradigm, one stressor is applied per day for a total of 28 consecutive days (Fig. 1). For the predator odor stressor, mice are placed in the rat home cages for 15 h (6 p.m. to 9 a.m.). For the space restraint stressor, mice are placed for a period of 2 h into the well-ventilated transparent tube described previously. Mice are subjected to tail suspension stress for 6 min as described above.

(B) *SPT*

Anhedonia is evaluated by the SPT before starting the CMS paradigm (day 0) and after its termination (day 29). For the first SPT, mice are subjected to a *habituation phase*, in which they are familiarized with 2% sucrose solution, before the actual *test phase*. The latter evaluates the preference for the sweet solution versus water by comparing the consumed volumes in a two bottle-forced choice paradigm.

The habituation phase lasts 3 days. At day 1, food and water is removed for a deprivation period of 18 h starting at 3 p.m. At day 2, after the 18 h deprivation period (at 9 a.m.), food is restored and one customized drinking bottle (described above) filled with 2% sucrose solution is placed into one side of the wire lid for 24 h. To prevent possible effects of side preference in drinking behavior, the position of the bottle is changed to the other side of the wire lid after 24 h (at 9 a.m. of day 3). On day 4 (at 9 a.m.), the bottle containing the sucrose is removed and regular food and water is restored for 6 h. During the *test phase*, food and water is removed for a deprivation period of 18 h starting at 3 p.m. The following morning, customized drinking bottles are filled with either 2% sucrose solution or water and labeled properly for unambiguous identification. To avoid taste bias, the sucrose solution is prepared using the same type of water used for filling the customized water bottles (i.e., autoclaved water, if this is what is regularly used as drinking water and for filling the water bottles in the SPT) (*Note 6*).

Bottles are filled to the very top and kept bottom down. In the next step the bottles are weighed and their weight is recorded. The bottles are carefully introduced through the wire lids and are checked for dripping. It is important to verify that the sipper tubes fit through the wire lid far to allow for easy access by the mice (*Note 7*).

The actual testing period lasts for 3 h starting at 9 a.m. The exact time when the drinking bottle-containing wire lids have been placed into the cage is noted individually and mice remain undisturbed during the entire testing period. When the 3 h testing period is over, bottles are weighed again and the weight is noted for later analysis.

(C) *Assessment of fur state*

To assess levels of stress induced by the CMS procedure, the physical state of the fur is analyzed as indicator of well-being. The fur state is examined at 9 am on day 0, 7, 14, 21, and 29 of the CMS paradigm and scored according to the 4-points scale described above.

(D) Assessment of body weight

During the 28 days of the CMS paradigm, the body weight of the mice is evaluated weekly at a designated day and time.

3.2 Time Required

(A) *CMS paradigm*

The CMS paradigm is conducted over 28 days (Fig. 1). For the exposure to the predator stressor 15 h are required. Approximately 15 min need to be considered as preparation time of the mouse boxes destined for rat exposure stress. The time necessary for the restraint stressor is 2 h per day. Preparation time for each tube is approximately 5 min. The tail suspension requires 6 min per day. Twenty to thirty minutes of additional time is required before the start of the stressor for preparation of equipment, transport of animals to the rat housing room, etc. Approximately 1–2 min per mouse is needed to place the first aid tape on the tip of the mouse tail and to remove it at the end of each session.

(B) *SPT*

For the initial SPT 5 days are required, as a *habituation phase* (3 days) needs to precede the *testing phase* (see above for details). For the second SPT (after CMS) the *habituation phase* is omitted and the actual testing phase begins with food restriction the day before the 3 h choice phase the following morning (as described before). Thirty minutes are required to initially assemble and fill the bottles. Please refer to Table 1 for the 5 day time frame. The time necessary for subsequent data analysis (described below) depends on the size and complexity of the experiment.

(C) *Assessment of fur state and body weight*

Fur state and body weight can be monitored at the same time. On average 2 min for each animal are required.

3.3 Data Analysis

(A) *SPT*

Sucrose preference rate is calculated as the amount of sucrose solution relative to the total fluid intake using the following formula:

$$\text{Sucrose preference rate} = \text{sucrose solution intake}(g) / \left[\text{sucrose solution intake}(g) + \text{water intake}(g)\right] \times 100\%$$

Table 1

Temporal course of the sucrose preference test

Day of experiment	Duration (h)	Explanation
1	18	Food and water deprivation
2	48	2% sucrose water introduction and food restoration
4	18	Food and water deprivation
5	3	Sucrose preference test

Differences in sucrose preference between two groups (CMS paradigm, drug treatment, mouse strain, genetic manipulation etc.) are evaluated for statistically significant differences using Student's-*t*-Test or Mann-Whitney-*U*-Test, if the data violate the principle of parametric distribution. ANOVA-based statistical evaluation is employed when comparing more than two groups (one-way) or more than one condition (two or three-way). A minimal number of ten animals per group is estimated to be required for meaningful statistical analysis based upon own experience as well as effects described in the literature. The actual number is calculated using a power analysis of the expected changes of the main experimental parameters (http://www.quantitativeskills.com/sisa/).

(B) *Fur state*

For each mouse the difference between the score assigned the day before starting the CMS paradigm (day 0) and the first day at the end of the paradigm (day 29) is calculated. The mean difference of fur scores for each experimental group is statistically examined as described above.

(C) *Body weight*

Body weight loss/gain is calculated as relative amounts of body weight loss at the end of the CMS paradigm (day 29) with respect to the initial body weight (day 0) using the following formula:

$$\text{Body weight change} = \left[\begin{pmatrix} \text{Body weight day}(0) - \\ \text{Body weight day}(29) \end{pmatrix} / \text{ Body weight day}(0) \right] \times 100\%$$

Statistical evaluation is carried out similar to the procedure described for the SPT.

4 Experimental Variables

(A) *CMS paradigm*

Chronic exposure to a series of mild stressors has been shown to decrease hedonic responses in several strains of laboratory rats and mice [8, 11, 20, 21]. Although the anhedonia-inducing effects of the CMS procedure are generally observable across different species and strains within these individual species, it must be pointed out that some strains of rats and mice are less susceptible to CMS (e.g., Flinders Resistant Line (FRL) which are less susceptible than the FSL strain [21]). Similarly, DBA/2 mice are less susceptible than CBA/H and C57BL/6 mice [20]. Hence, the genetic, background, age and sex [22] need to be taken into account when designing and evaluating the behavioral outcome of the CMS paradigm.

(B) *SPT*

The SPT has some advantages over other tests used to evaluate reward-related behaviors: it can be performed without cost-intensive apparatus and it is non-invasive. Furthermore, it permits the evaluation of progressive states of depression or the time course of a treatment when repeated measurements need to be taken [23].

Inbred mouse strains have been shown to vary substantially in their baseline sucrose preference using other protocols [20]. Thus, adaptations of the presented protocol to a different mouse strain may be required to prevent a potential interaction of the genetic background with the experimental manipulation tested.

It has been shown that the CMS paradigm induces anhedonia only in a subgroup of susceptible animals, termed "anhedonic mice" [15]. Animals that have undergone CMS procedures and do not show hedonic deficits are resilient and are usually indicated as "non-anhedonic mice" and can be used as internal control for unspecific effects of stressor exposure which is not directly associated with anhedonia [15, 24]. It has been argued that decreases in hedonic responsiveness following CMS may be caused by changes in body weight resulting from the CMS procedure [16]. However a large number of studies showed that hedonic sensitivity relative to body weight (e.g., mg sucrose/g body weight) was decreased significantly by CMS [8] suggesting that decreased hedonia following CMS cannot be attributed to loss of body weight, but rather to a decreased hedonic sensitivity.

(C) *Fur state*

Sex differences and strain-specific patterns of grooming behavior have been reported in mice [25] and may constitute potential confounding factors for the analysis of fur state.

(D) *Body weight*

Among various phenotypes associated with depression, decreased anhedonic responsiveness and changes in body weight seem occur independently of each other [8, 15]. The range of body weight loss typically observed in CMS experiments may vary between different laboratories or even between different studies conducted in the same laboratory [26, 27]. It has been argued that the extent of weight loss can be caused by periods of food and water deprivation during different CMS paradigm used [28]. However, in the majority of the studies, a weight decrease was also observed in CMS protocols which excluded food or water deprivation [29]. Similar patterns of body weight fluctuations have been observed in clinical settings where, MDD patients report a gain/reduction in weight. On average patients tend to lose approximately 7% of body weight in the course of the depressive illness [30]. An equivalent effect is seen in rodent models of depression: the age-dependent weight gain is blunted (see below).

5 Anticipated Results

(A) *SPT*

Depression-like behavior in anhedonic mice is inferred from the significant decrease in sucrose preference from the initial SPT (before CMS) to the SPT analysis conducted after the CMS procedure. A representative result from our laboratory is depicted in Fig. 2. We typically find ~50% of animals to be resilient to CMS induced anhedonia—a response rate comparable to previously obtained effects described as partial depressive-like outcomes in chronic stress models of depression in the literature [15, 31, 32]. Different approaches have been described to define the effect, i.e., susceptibility corresponding to the induction of chronic mild stress-induced anhedonia. Some studies rely on an absolute "cut-off" to define the state of anhedonia, e.g., sucrose preference of ≤65% [15]. Alternatively, a relative approach can be used which allows for taking into account individual baseline differences in sucrose preference: in this approach, the change in sucrose preference after the CMS procedure (or handled-only control) is related to the individual's baseline value before the start of CMS. The average percentage of change in control (non-stressed animals, handled daily throughout the CMS period) animals is used as criterion to separate responders (susceptible) from non-responders (resilient).

(B) *Fur state*

A drop in grooming and consequently a higher fur state score can be expected in animals exposed to the CMS procedure, but

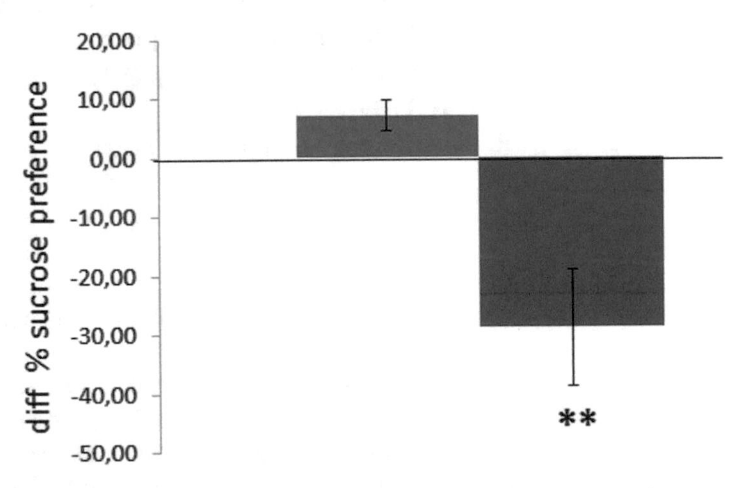

Fig. 2 Change in sucrose preference induced by chronic mild stress exposure in anhedonic mice. CMS exposure induces a decrease in the percentage of sucrose preferences in anhedonic CMS exposed (*red*) but not in control (handled-only) male adult C57Bl6/N mice ($n = 8$–11 per group). Data are displayed as mean ± SEM, $^{**}P < 0.01$

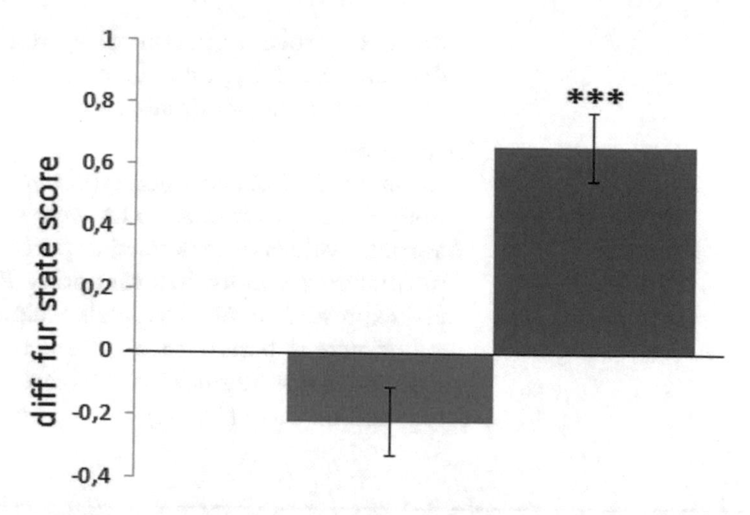

Fig. 3 Physiological body weight gain is altered in mice after chronic mild stress. CMS exposed (*red*) male adult C57Bl6/N mice show a decrease in the percentage of body weight change during the course of the experiment while body weight increases in control (handled-only) animals ($n = 8$–11 per group). Data are displayed as mean ± SEM, ***$P < 0.001$

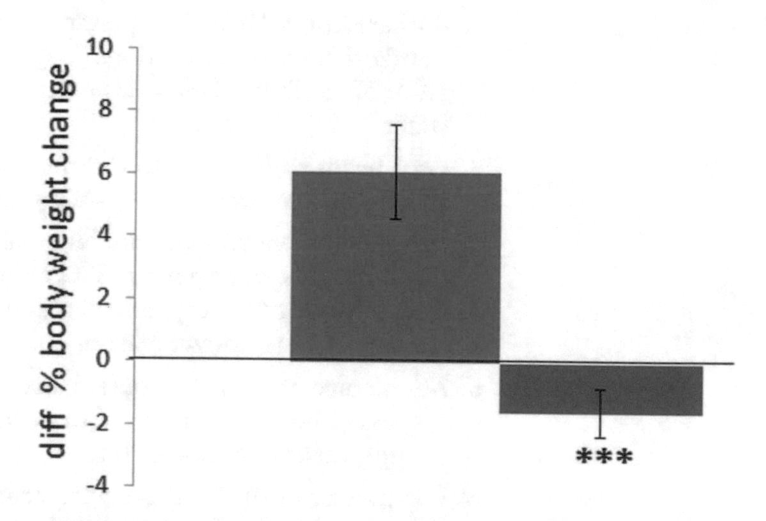

Fig. 4 Deterioration of the fur state resulting from exposure to chronic mild stress. CMS exposed (*red*), but not control (handled-only) male adult C57Bl6/N mice animals present with deterioration of the fur state over the experimental period group ($n = 8$–11 per group). Data are displayed as mean ± SEM, ***$P < 0.001$

not in controls, when comparing the score before and after the 4-week period. Figure 3 illustrates the result from a representative experiment of our laboratory.

(C) *Body weight*

Adult single-housed mice typically show a steady increase in body weight over time. The weight gain can be expressed in absolute values or expressed as percent of initial weight, which normalizes for individual variations. Regardless of how the data are expressed, control animals weigh more after the 4 weeks experimental period than at the beginning. However, body mass does not augment or increases to a lesser extent in mice exposed to CMS (Fig. 4).

6 Notes

1. Drug administration must only be carried out by trained personnel for safety reasons, to reduce discomfort for the experimental animals and to ensure accurate drug delivery. When injections are administered, used needles are disposed properly in designated receptacles and each needle is discarded after single use.
2. Rats are housed in an independent experimental room, under standard lighting conditions (approximately 200 lux) on a 12 h:12 h light–dark cycle with free access to food and water.
3. Care needs to be taken to put the mouse in the center of the scale to prevent improper fluctuations of the weight.
4. It is recommended that handling and behavioral tests are carried out by the same person. This allows the mice to familiarize themselves with the experimenter and eliminates the additional stressor of an unknown person.
5. Mice naïve to human contact may react with defensive or aggressive behaviors and need to be handled with caution, preventing them to escape or bite.
6. For validity of the SPT, accurate determination of liquid consumption is pivotal hence dripping of the bottles has to be prevented. Since changes in temperature may lead to leakage of bottles, they are prepared and kept at the same ambient temperature of the room in which the SPT is carried out.
7. Positions (right or left side) of the sucrose bottles are alternated within the experimental groups to avoid effects of side preference.

7 Conclusions

Major depressive disorder is a complex and multifaceted human mental illness. Excellent reviews are available, which describe the relevance and limitations of animal models in depression research; the reader is referred to this literature [3–5]. It is obvious that (i) each model can only address some aspects of the human disease, (ii) the interpretation must be based on a solid background in the underlying neurobiology, and (iii) the power of a given approach is also limited by the experimental variability. The latter can be reduced by attention to detail. The CMS model has several advantages over other attempts to mimic elements of depression in animals. This includes the inexpensive setup, which allows for establishing this paradigm without major investment into equipment. Importantly, the microstressors applied are ethically sound and do not physically harm the mice in any way. Finally, the CMS procedure allows for evaluating evolution of physical and behavioral changes over time, thus enabling also a within-subject control design. The SPT provides for a robust method to evaluate anhedonic behavior. However, it is important to keep in mind that anhedonia is only one symptom of depression. Evaluation of other parameters—such as changes in fur state and body weight—may provide additional insights. It still remains incompletely understood, why some animals are susceptible or resilient to CMS. However, this is also true for human depression, where equivalent stressors produce very different outcomes. The biological basis for susceptibility and resilience may—at least in part—be shared by mice and men [33]. Hence, although no model is perfect and no test infallible, behavioral paradigms will remain important tools in the foreseeable future to enhance our understanding of the pathophysiological underpinnings of multifaceted psychiatric disorders such as depression and anxiety.

Acknowledgements

The authors wish to thank the Austrian Science Fund for continuous support (grant F35).

References

1. Bromet E, Andrade LH, Hwang I et al (2011) Cross-national epidemiology of DSM-IV major depressive episode. BMC Med 9:90
2. Alesci S, Martinez PE, Kelkar S et al (2005) Major depression is associated with significant diurnal elevations in plasma interleukin-6 levels, a shift of its circadian rhythm, and loss of physiological complexity in its secretion: clinical implications. J Clin Endocrinol Metab 90:2522–2530
3. Cryan JF, Slattery DA (2007) Animal models of mood disorders: recent developments. Curr Opin Psychiatry 20:1–7
4. Pollak DD, Rey CE, Monje FJ (2010) Rodent models in depression research: classical strategies and new directions. Ann Med 42:252–264
5. Rupniak NM (2003) Animal models of depression: challenges from a drug development perspective. Behav Pharmacol 14:385–390

6. Bagot RC, Labonte B, Pena CJ et al (2014) Epigenetic signaling in psychiatric disorders: stress and depression. Dialogues Clin Neurosci 16:281–295
7. Slavich GM, Irwin MR (2014) From stress to inflammation and major depressive disorder: a social signal transduction theory of depression. Psychol Bull 140:774–815
8. Willner P, Towell A, Sampson D et al (1987) Reduction of sucrose preference by chronic unpredictable mild stress, and its restoration by a tricyclic antidepressant. Psychopharmacology (Berl) 93:358–364
9. Lucca G, Comim CM, Valvassori SS et al (2008) Chronic mild stress paradigm reduces sweet food intake in rats without affecting brain derived neurotrophic factor protein levels. Curr Neurovasc Res 5:207–213
10. Martin M, Ledent C, Parmentier M et al (2002) Involvement of CB1 cannabinoid receptors in emotional behaviour. Psychopharmacology (Berl) 159:379–387
11. Monleon S, D'aquila P, Parra A et al (1995) Attenuation of sucrose consumption in mice by chronic mild stress and its restoration by imipramine. Psychopharmacology (Berl) 117:453–457
12. Valverde O, Smadja C, Roques BP et al (1997) The attenuation of morphine-conditioned place preference following chronic mild stress is reversed by a CCKB receptor antagonist. Psychopharmacology (Berl) 131:79–85
13. Azpiroz A, Fano E, Garmendia L et al (1999) Effects of chronic mild stress (CMS) and imipramine administration, on spleen mononuclear cell proliferative response, serum corticosterone level and brain norepinephrine content in male mice. Psychoneuroendocrinology 24:345–361
14. Savalli G, Diao W, Schulz S et al (2015) Diurnal oscillation of amygdala clock gene expression and loss of synchrony in a mouse model of depression. Int J Neuropsychopharmacol 18:pii: pyu095
15. Strekalova T, Spanagel R, Bartsch D et al (2004) Stress-induced anhedonia in mice is associated with deficits in forced swimming and exploration. Neuropsychopharmacology 29:2007–2017
16. Forbes NF, Stewart CA, Matthews K et al (1996) Chronic mild stress and sucrose consumption: validity as a model of depression. Physiol Behav 60:1481–1484
17. Papp M, Willner P, Muscat R (1991) An animal model of anhedonia: attenuation of sucrose consumption and place preference conditioning by chronic unpredictable mild stress. Psychopharmacology (Berl) 104:255–259
18. Mineur YS, Belzung C, Crusio WE (2006) Effects of unpredictable chronic mild stress on anxiety and depression-like behavior in mice. Behav Brain Res 175:43–50
19. Mineur YS, Prasol DJ, Belzung C et al (2003) Agonistic behavior and unpredictable chronic mild stress in mice. Behav Genet 33:513–519
20. Pothion S, Bizot JC, Trovero F et al (2004) Strain differences in sucrose preference and in the consequences of unpredictable chronic mild stress. Behav Brain Res 155:135–146
21. Pucilowski O, Overstreet DH, Rezvani AH et al (1993) Chronic mild stress-induced anhedonia: greater effect in a genetic rat model of depression. Physiol Behav 54:1215–1220
22. Franceschelli A, Herchick S, Thelen C et al (2014) Sex differences in the chronic mild stress model of depression. Behav Pharmacol 25:372–383
23. Schweizer MC, Henniger MS, Sillaber I (2009) Chronic mild stress (CMS) in mice: of anhedonia, 'anomalous anxiolysis' and activity. PLoS One 4:e4326
24. Tang M, Lei J, Sun X et al (2013) Stress-induced anhedonia correlates with lower hippocampal serotonin transporter protein expression. Brain Res 1513:127–134
25. Kalueff AV, Minasyan A, Keisala T et al (2006) Hair barbering in mice: implications for neurobehavioural research. Behav Processes 71:8–15
26. Cheeta S, Broekkamp C, Willner P (1994) Stereospecific reversal of stress-induced anhedonia by mianserin and its (+)-enantiomer. Psychopharmacology (Berl) 116:523–528
27. Cheeta S, Ruigt G, Van Proosdij J et al (1997) Changes in sleep architecture following chronic mild stress. Biol Psychiatry 41:419–427
28. Matthews K, Forbes N, Reid IC (1995) Sucrose consumption as an hedonic measure following chronic unpredictable mild stress. Physiol Behav 57:241–248
29. Muscat R, Papp M, Willner P (1992) Antidepressant-like effects of dopamine agonists in an animal model of depression. Biol Psychiatry 31:937–946
30. Weissenburger J, Rush AJ, Giles DE et al (1986) Weight change in depression. Psychiatry Res 17:275–283
31. Bergstrom A, Jayatissa MN, Thykjaer T et al (2007) Molecular pathways associated with stress resilience and drug resistance in the chronic mild stress rat model of depression: a gene expression study. J Mol Neurosci 33:201–215
32. Vialou V, Robison AJ, Laplant QC et al (2010) DeltaFosB in brain reward circuits mediates resilience to stress and antidepressant responses. Nat Neurosci 13:745–752
33. Mcguffin P, Alsabban S, Uher R (2011) The truth about genetic variation in the serotonin transporter gene and response to stress and medication. Br J Psychiatry 198:424–427

Chapter 12

In Vivo Imaging of Dopamine Metabolism and Dopamine Transporter Function in the Human Brain

Matthäus Willeit, Ana Popovic, Lucie Bartova, Ulrich Sauerzopf, Martin Bauer, and Nicole Praschak-Rieder

Abstract

Positron emission tomography (PET) and single-photon emission computed tomography (SPECT) imaging of the dopamine system allow quantifying specific targets in the living animal and human brain. These methods are thus of great importance for translational brain research and have made it possible to identify and measure neurochemical changes associated with psychiatric disorders for the first time in history. The following chapter focuses on PET and SPECT imaging of psychotic disorders and addresses methods suited for imaging changes in extracellular dopamine levels and their relationship to dopamine metabolism and dopamine transporter function. Specifically, the chapter describes imaging with radiolabeled dopamine precursors (such as [^{18}F]DOPA) and the so-called "competition paradigms," where a change in extracellular dopamine elicits changes in radioligand binding to dopamine $D_{2/3}$ receptors. In addition to theoretical background, this chapter provides information on strengths and weaknesses as well as on practical aspects of these methods.

Key words Dopamine, DOPA, Psychosis, Schizophrenia, Amphetamine, PET, SPECT, [^{11}C]-(+)-PHNO

1 Introduction

The dopamine transporter (DAT) is an essential element in the regulation of intensity, duration, and spatial spread of brain dopamine signalling. DAT is the target of therapeutic and abused drugs such as methylphenidate, amphetamines, and cocaine. Imaging reductions in radioligand binding to DATs in the living brain using single-photon emission computed tomography (SPECT) or positron emission tomography (PET) has become an essential tool for aiding clinicians in diagnosing Parkinson's disease and other neurodegenerative disorders involving brain dopamine functions. Moreover, changes in DAT binding have been observed in populations with psychiatric disorders such as addiction or attention-deficit hyperactivity disorder (ADHD). The methods used for

Heinz Bönisch and Harald H. Sitte (eds.), *Neurotransmitter Transporters: Investigative Methods*, Neuromethods, vol. 118, DOI 10.1007/978-1-4939-3765-3_12, © Springer Science+Business Media New York 2016

DAT imaging can be divided into those aiming at quantifying the amount of DAT protein accessible to radioligand binding, and those assessing DAT and dopamine function and metabolism. While quantifying DAT binding using PET or SPECT is comparably straightforward, measurement of DAT function in the living brain requires a more complex approach, and the interpretation of data relies on several assumptions that need to be tested and verified in animal experiments in vitro and in vivo.

Essentially, two different approaches have been used to study DAT function in the living human brain. One is using a radiolabeled version of the dopamine precursor dihydroxy-phenylalanine (DOPA), most commonly [^{18}F]DOPA. The other approach is to study the interaction of endogenous extracellular dopamine with postsynaptic receptors, most commonly dopamine D_2 and D_3 receptor subtypes. For this technique, a behavioural or pharmacologic intervention is used to alter extracellular dopamine levels. Typically, an increase in extracellular dopamine leads to reductions in radioligand binding, while a decrease in extracellular dopamine is associated with increased radioligand binding. This allows for an estimate of changes in extracellular dopamine levels in the living human brain. Although imprecise, the term "competition" or "displacement" study is frequently used to denote this method (see below for further discussion of this topic).

Both methods have provided significant insight into the pathogenesis of psychiatric disorders, first and foremost into altered dopamine transmission in psychosis or schizophrenia. In these conditions, macroscopic alterations of the brain—if present at all—are subtle or visible only after many years of active illness. Thus, with the notable exception of helping in the differential diagnosis to the so-called organic brain disorders, structural imaging methods have only limited diagnostic or prognostic relevance for clinical practice in the field of psychotic disorders. And as of yet, structural imaging has contributed little to our pathogenic understanding of these disorders.

In contrast, PET imaging of the dopamine system has clearly shown that dopamine function is altered in psychosis or schizophrenia. Dopamine synthesis and storage, as well as amphetamine-induced release of dopamine into the extracellular space are increased in psychotic patients. Moreover, studies in the prodromal phase of the illness have shown that alterations in brain dopamine function are detectable even before subjects develop clinical symptoms of psychosis [1, 2]. Thus, although costly and so far available to few specialized centers only, imaging methods based on the aforementioned techniques have the potential to become a useful clinical tool in the foreseeable future. Imaging could support clinicians in diagnostic and prognostic assessments or help stratifying patients for choosing specific treatment modalities.

Here, we try to provide a concise overview on basic principles of PET imaging and on possibilities and limits of studying dopamine metabolism and DAT function in the living human brain. This chapter is written for all those who are interested in the results of, but are not directly involved into PET studies on the dopamine system in the living human brain, or who think they may profit in one or the other way from some methodological background information.

2 Materials and Methods

2.1 Imaging Procedures

Radioligands are produced in a complex and resource-intensive process using a cyclotron for generating positron-emitting isotopes. For PET ligands used in studies of DAT function, the isotope is, at the time being, either flourine-18 ([^{18}F]) or carbon-11 ([^{11}C]). A practical advantage of [^{18}F]-labeled radiotracers over [^{11}C]-labeled ones is the slower decay (half-life of [^{18}F]: 109.8 min; [^{11}C]: 20.3 min), allowing for storage and transport for a limited time period. [^{11}C]-labeled ligands need to be produced on-site in close proximity to the PET scanning system. Target specificity and affinities to dopamine and non-dopamine receptors, brain penetration, protein binding, lipophilicity, metabolic stability and many other parameters need to be characterized extensively in vitro and in animal models before a radioligand is newly introduced to human PET imaging (see for example [3]). For a PET scan, the radioligand is injected as a bolus into a peripheral vein, usually over a time period of approximately 1 min. However, alternative methods exist, for example bolus plus constant infusion paradigms discussed later in this chapter. Depending on the radioligand and the imaging protocol, subjects need to lie supine in the scanner for 60–90 min, and sometimes for up to 3 h. The effective radiation dose injected is determined by the radioligand and its specific activity and usually ranges from less than 1 to up to 5 mSv (the average radiation dose from natural environmental sources to humans worldwide is approximately 2.4 mSv per year). Exposure to radioactivity is the limiting factor for the number of PET scans that can be performed in a subject for the purpose of research. A coincidence-detector within the PET camera is then used to detect and localize positron-annihilation events in the brain. The output of the so-called data reconstruction provides a temporo-spatial count-matrix into a four-dimensional image as output. During this process, data are corrected for heterogeneities in signal-attenuation brought about by anatomical variations in tissue composition. This correction is performed using an attenuation-matrix obtained in a so-called transmission scan performed directly before or after each PET scan. Furthermore, correction for radioactive decay is applied. Current high-resolution PET imaging systems can ascribe decay-events to cubic volume units (voxels) with a side-length of 1.25 mm. However, this resolution is an ideal value reached at the center of the field-of-view only, and the reliability of the information depends on a variety of factors, such as image-reconstruction algorithms, partial volume effects, movement artefacts and so on. For improving anatomical accuracy of measurements, a magnetic resonance image (MRI) is acquired in a separate scanner and then "merged" with the PET image at a later point in time (see below). Combined PET/MRI imaging systems allowing for simultaneous MRI and PET measurements are a recent technological innovation available to few specialized research facilities only.

2.2 Image Analysis

PET data are analyzed using a broad variety of methods. Here, we confine the description to methods most commonly used for analysis of [^{18}F]DOPA and $D_{2/3}$ "competition" data. The fundamental principle of neuroreceptor (understood as general term including also DAT) PET is to use pharmacokinetic information acquired during a PET scan for deriving a set of parameters describing ligand–receptor interaction in the brain.

2.3 Preprocessing of Imaging Data

Procedures of PET image analysis may vary greatly according to PET and MRI systems used, anatomical targets structures, radioligands, and specific study aims. Here we describe procedures that are frequently used for DAT or dopamine $D_{2/3}$ receptor imaging. Usually, PET and MRI images are co-registered using linear iterative interpolation algorithms implemented into freely or commercially available software packages (e.g., Statistical Parametric Mapping; SPM). In study protocols using repeated scans, this step also allows to bring all image-sets of one subject into the same space. Usually, this intermodal brain image registration is attained using the so-called "rigid-body co-registration" algorithms, in which images, in order to yield the best possible overlay, are moved in space but are not deformed.

Another approach is used for direct voxel-wise comparisons of PET data from different subjects. First PET and MRI images are rigid-body co-registered and then transformed ("normalized") into a standard space, for example "Montreal Neurologic Institute (MNI) space." A standard space is an anatomical template created by averaging many individual images, usually MRI images. In standard space, anatomical structures have fixed coordinates allowing for comparison of results independent of study equipment and individual anatomical variations. The rationale for using MRI images for these steps is that MRI images contain much more and more accurate anatomical information than PET images. The transformation matrix is then used to normalize the respective PET images. This approach is widely used for processing parametric PET data (parametric maps), when a tracer kinetic model is fit to time–activity curve data on a pixel-by-pixel basis. The result is a quantitative image of the parameters of a physiological or biochemical model.

2.4 Region-of-Interest Analysis

Tissue composition, receptor population and other parameters of relevance vary greatly within the brain. However, the analysis of PET data makes the assumption of uniform conditions within a given volume. Traditionally, these volumes are denoted "region of interest" (ROI). ROIs are selected according to anatomical and functional criteria and are usually brain regions where the respective radioligand shows sufficient and reversible (within the time of a PET scan) binding. In imaging DAT function, ROIs usually comprise DAT and $D_{2/3}$ receptor rich regions such as the striatum and brainstem regions. In studies using PET cameras with high resolution, it is possible to image subdivisions of the striatum such as the ventral striatum (VST), putamen (PUT), caudate nucleus (CAU), and globus pallidus (GP). Resolution of current PET systems does usually not allow for a reliable distinction between the substantia nigra (SN) and the ventral tegmental area (VTA). Thus, these areas are usually analyzed in a combined brainstem ROI (SN/VTA). Moreover, high-resolution PET systems have lately been able to detect specific binding in small regions with only intermediate levels of dopaminergic target molecules, such as hippocampus or amygdala. For the analysis, decay-corrected regional radioactivity is used to measure pharmacokinetic behaviour of the radioligand in a given ROI. Graphically, this gives the so-called "time–activity curves" (TACs). These curves are then analyzed using a variety of methods, most frequently by iterative fitting to predefined compartmental models.

2.5 Time–Activity Curves (TACs) and Analysis of Binding Parameters

The primary outcome measures in neuroreceptor PET imaging are TACs. TACs are obtained by plotting decay-corrected regional radioactivity against time (see Fig. 1 for exemplary TACs obtained with the dopamine D2/3 receptor agonist radioligand [^{11}C]-(+)4-propyl-9-hydroxynaphthoxazine; [^{11}C]-(+)-PHNO; Fig. 2). The pharmacokinetic behaviour of the radioligand in a given ROI is used to derive a set of binding parameters describing pharmacological properties of the ligand–receptor interaction such as maximal binding capacity (B_{max}), affinity (given by the dissociation constant K_d), or, in case of [^{18}F]DOPA PET, the influx constant K_i.

Parameters are estimated by iterative fitting of PET data to a mathematical model known to adequately reflect pharmacokinetics and binding behaviour of the respective radioligand in the brain. Depending on the ligand, models need to take into account the exchange of radioligand between various tissue compartments, formation of radioactive metabolites and their diffusion or active transport between compartments, plasma protein binding etc. A scheme representing a so-called three-compartment model is depicted in Fig. 3. This model reflects imaging protocols using a so-called arterial input function, where blood is repeatedly collected from a peripheral artery (in most cases an artery at the wrist).

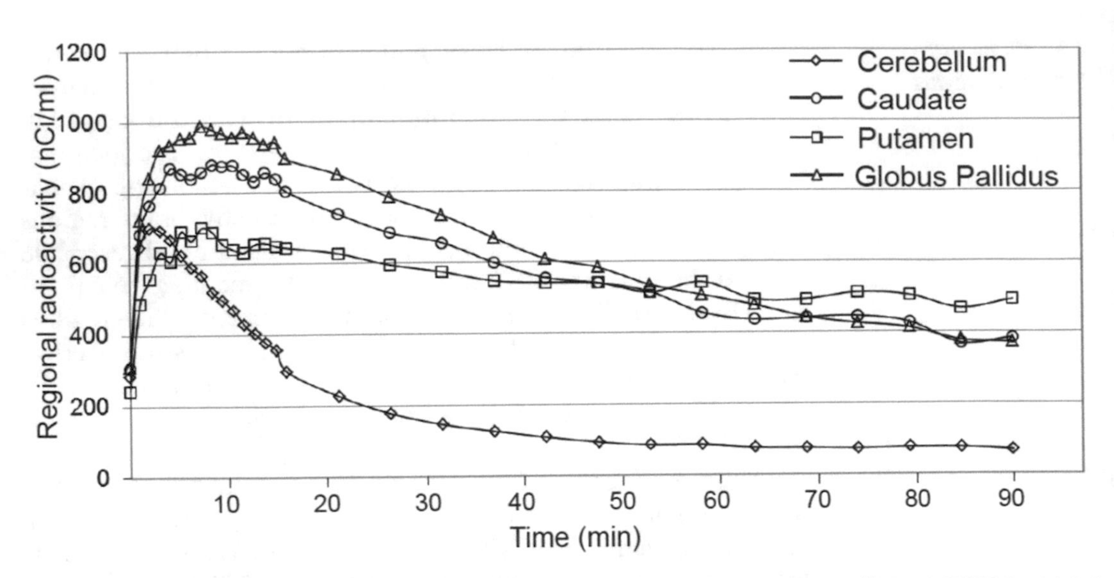

Fig. 1 Time–activity curves (TACs) for the dopamine $D_{2/3}$ receptor agonist-radioligand [^{11}C]-(+)-PHNO in striatal regions of interest (ROIs) and the cerebellum (reference region) in a healthy human subject

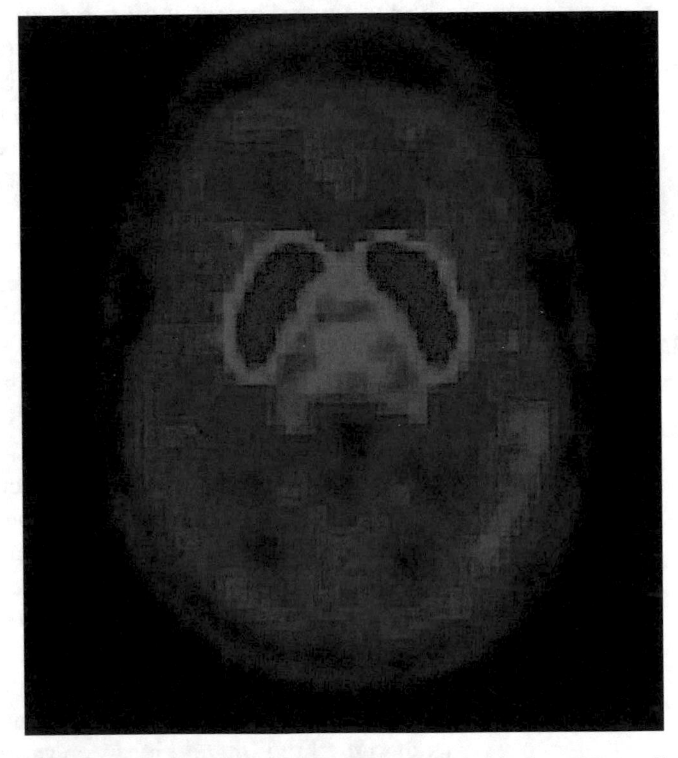

Fig. 2 Average image of a dynamic PET scan showing binding of the radioligand [^{11}C]-(+)-PHNO to dopamine $D_{2/3}$ receptors in the human striatum

Blood is analyzed for total and protein-bound radioligand concentration for obtaining free plasma radioligand concentration. Free plasma concentration is assumed being the true concentration delivered through the blood–brain barrier.

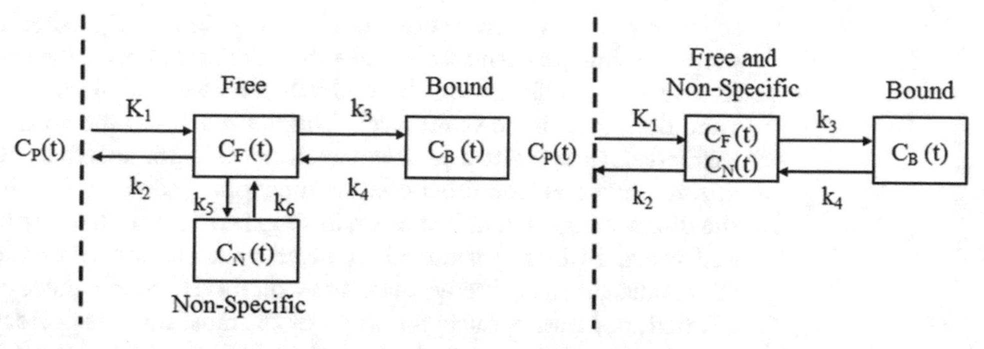

Fig. 3 Schematic of a three- (*left*) and two-compartment (*right*) model describing movement of a radioligand between compartments. The free radioligand fraction can be distinguished from nonspecific binding by using an arterial input function. This is not possible for a noninvasive reference region approach. These fractions are thus treated as one compartment assuming equal conditions in receptor-rich and reference regions. K_1 denotes the rate constant for transfer from plasma to tissue (unit: mL cm^{-3} min^{-1}). k_2, k_3, k_4, k_5, and k_6 denote rate constants for transfer into or out of the compartments as shown in the diagram (unit: min^{-1}). k_3 and k_4 reflect the ligand–receptor association (k_{on}) and dissociation (k_{off}) rate constants at a molecular level. And their ratio (k_4/k_3) is proportional to the in vitro dissociation constant K_d

2.6 Analysis Using Noninvasive Reference-Region Approaches

Since collection and analysis of arterial blood is somewhat painful for the imaged subject and cumbersome for the research team, noninvasive reference tissue models [4] are applied whenever possible. The PET signal in the brain reflects a mixture of free (unbound), nonspecifically bound, and specifically bound radioligand. A reference region is a brain region that contains no or negligible amounts of the targeted receptors (here DAT or $D_{2/3}$ receptors) but has a similar tissue composition and thus, comparable nonspecific binding. Free and nonspecific binding is collapsed into one compartment, and specific binding in the ROI is calculated using rate constants for free and nonspecific binding measured in the reference region. Feasibility and specifics of a reference tissue approach are determined for every radioligand in what is commonly called "modeling" of a radioligand. For the purpose of modeling, data obtained with an arterial input function (see for example [5, 6]) are used as a gold standard.

For dopamine imaging, the most commonly used reference regions are the cerebellar or occipital cortex, as they contain negligible dopaminergic innervation. The divergent pharmacokinetics of the $D_{2/3}$ receptor radioligand [^{11}C]-(+)-PHNO in $D_{2/3}$ receptor-rich striatal regions and the cerebellum can be easily recognized in Fig. 1. With respect to the cerebellum, it is an important detail to use cortical cerebellum with a certain distance to midline structures such as the vermis, as these midline structures have been shown to exhibit some degree of specific binding.

2.7 Parametric Analysis

The so-called "parametric" analysis methods treat every voxel in a PET image as ROI, i.e., kinetics of radioligand binding are separately analyzed in every voxel in the brain [7]. After data preprocessing (anatomical normalization etc., see above), the behaviour of the radioligand can be compared between subjects and/or conditions at a single-voxel level. This approach has the advantage that it does not require a-priori an anatomical hypothesis that might, at times, conceal true functional connections. However, it conveys the disadvantage that information in single voxels is inherently noisy and prone to bias introduced by artefacts such as partial volume effects and others. Moreover, statistical methods become more complicated, not least because the number of statistical tests performed in a single analysis equals the number of voxels in the image. The methods for addressing the multiple-comparisons problem and other peculiarities are similar to those applied in functional magnetic resonance imaging (fMRI), and usually, parametric PET maps are analyzed using adapted software originally designed for fMRI. A frequently adopted approach is to combine parametric and traditional ROI analysis methods, e.g., by "masking" parametric maps in order to restrict the voxel-wise analysis to relevant brain areas.

2.7.1 [^{18}F]DOPA PET

After exogenous administration, DOPA is reversibly transported through the blood–brain barrier (BBB) by a transporter that DOPA is sharing with other amino acids [8]. This leads to equilibrium conditions between brain and plasma DOPA concentrations. Dopamine is formed from DOPA by decarboxylation via the enzyme aromatic amino-acid decarboxylase (AAADC, also known as DOPA-decarboxylase). The rate-limiting step in catecholamine synthesis under physiological conditions is hydroxylation of tyrosine via tyrosine hydroxylase. This step is no longer needed if DOPA is administered exogenously. Studies on [^{18}F]DOPA uptake and utilization thus depend on metabolic and transport events distal to tyrosine and tyrosine hydroxylation. Although AAADC is not specific for DOPA, its activity is highest in dopamine rich brain regions. Since dopamine no longer crosses the BBB, and since from a kinetic point of view, it is "trapped" in the brain compartment, AAADC plays a crucial role in the formation of the [^{18}F]DOPA signal in the brain. In neurons, dopamine transport between the extracellular and intracellular compartment depends primarily on DAT. Cytoplasmatic dopamine is transported into synaptic vesicles by the vesicular monoamine-transporter (VMAT), where it is stored and protected from degradation by monoamine oxidase.

Ultimately, the [^{18}F]DOPA signal reflects complex metabolic pathways in which DOPA is a precursor for epinephrine, norepinephrine, the so-called trace amines, or inert catabolic products. Consequently, DAT function is only one of several processes contributing to the [^{18}F]DOPA PET signal in the brain. DOPA (and [^{18}F]DOPA) is reversibly transported into an out of the brain (Fig. 4). A major metabolite of DOPA is *o*-methyl-dopamine (OMD). OMD is formed by methylation via catechol-*o*-methyltransferase (COMT), an

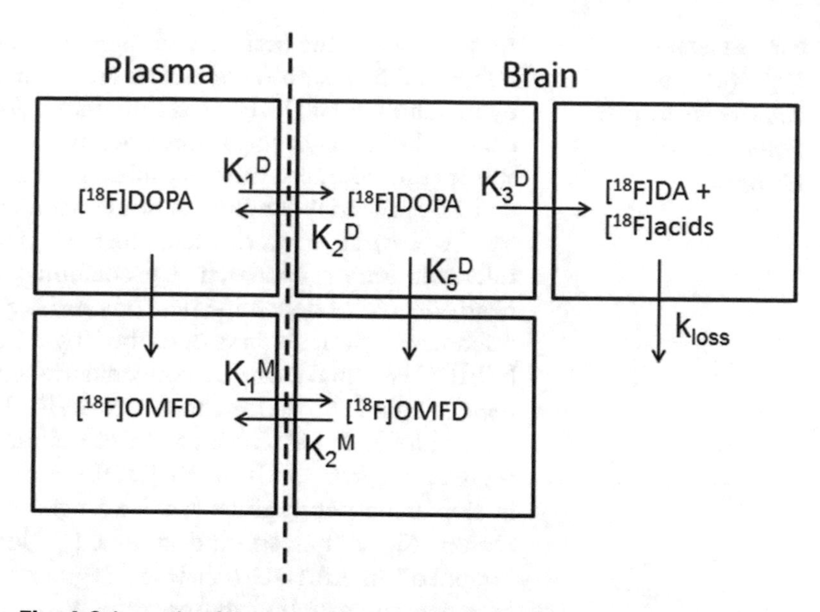

Fig. 4 Schematic of the compartmental model of [18F]dihydroxy-phenylalanine ([18F]DOPA) pharmacokinetics in the brain. [18F]DOPA is transported from blood plasma into (K_1^D) and out of the brain (K_2^D). In blood and brain, [18F]DOPA is metabolized by catechol-*o*-methyltransferase (COMT) to [18F]-o-methyl-fluorodopamine ([18F]OMFD) at the rate constant K_5^D. [18F]OMFD reversibly crosses the blood–brain barrier at the rates K_1^M and K_2^M. In dopaminergic neurons, [18F]DOPA is metabolized to [18F]dopamine by the enzyme aromatic amino-acid decarboxylase (AAADC) at the rate constant K_3^D, indicating the activity of AAADC relative to the concentration of its substrate. [18F]dopamine is stored in synaptic vesicles or metabolized to acidic degradation products leaving the brain by diffusion (k_{loss})

enzyme that inactivates dopamine in cortical brain regions where DAT expression is low. In the periphery, DOPA is metabolized by COMT in liver and blood cells. Thus, [18F]DOPA PET studies are usually carried out after administration of a COMT inhibitor such as carbidopa. Carbidopa does not enter the brain and prevents signal loss by peripheral degradation of [18F]DOPA. However, the presence of the [18F]-coupled OMD isotope ([18F]OMFD) in the brain, as well as the presence of radioactive dopamine degradation products should not be forgotten when interpreting [18F]DOPA PET results.

Using an arterial input function and correcting for [18F]DOPA metabolism, Cumming and colleagues [9] have shown that increased [18F]DOPA uptake can be present together with increased [18F]DOPA catabolism in patients with schizophrenia, indicating that the increase [18F]DOPA uptake- and storage-capacity found in patients with schizophrenia in several independent studies may co-occur with increased dopamine degradation. Using the same analysis method, it has been shown that [18F]DOPA turnover increases significantly after a single dose of methylphenidate [10].

2.8 Reference-Region Based and Linearized [18F] DOPA Analysis Methods

As seen above, the brain signal measured with [^{18}F]DOPA PET is influenced by a number of factors and composed to varying degree by metabolites of DOPA and dopamine. Although the interpretation of the signal is not straightforward, there is a large number of studies carried out in various clinical populations using [^{18}F]DOPA and PET. Arterial sampling is not convenient and at times, painful for the research subject. Thus, most [^{18}F]DOPA PET studies use a reference region approach. Cerebellum and occipital cortex are nearly devoid of dopamine neurons and are thus frequently used as reference region. As first described by Hartvig et al. 1993 [11], [^{18}F]DOPA uptake can be conveniently analyzed using linearized methods similar to a Scatchard plot or, for PET analysis, a Gjedde–Patlak plot [12–14]. In this form of analysis, the slope of the linear regression corresponds to [^{18}F]DOPA uptake (or AAADC activity) in dopamine rich regions relative to the input into the reference region. Since the assumption that [^{18}F]dopamine is irreversibly "trapped" in AAADC-rich brain regions is progressively violated by dopamine metabolism over time, this form of analysis is limited to a relatively brief period after tracer injection (usually about on hour). However, reference region based methods have successfully been used in many studies on Parkinson's disease or schizophrenia, and they have given reliable and replicated results reflecting reduced (Parkinson's disease) or enhanced (schizophrenia) dopaminergic neurotransmission in clinical populations.

In summary, the signal measured using [^{18}F]DOPA and PET is a complex composite measure of several enzymatic and transport processes involved in DOPA transport and metabolism, and dopamine synthesis, uptake, vesicular storage, catabolism, and subsequent elimination from the brain. In part, the relative contributions of these processes can be quantified when using arterial plasma sampling during image acquisition. Nevertheless, also approaches using more practicable reference region-based analysis methods have provided important insight in alterations of the dopamine system in clinical populations.

3 Competition Studies

Symptom provocation studies using DAT blockers or releasers are classical paradigms in the research on the biological basis of psychotic disorders (see [15] for review). Long before the mechanism of action of these drugs was understood, it was known that high doses of amphetamines are able to provoke psychotic symptoms in healthy subjects, and that patients with psychotic disorders show behavioural super-sensitivity towards amphetamines or methylphenidate (in this case, behavioral supersensitivity in patients indicates de novo occurrence or worsening of preexisting psychotic symptoms at doses of methylphenidate or amphetamine that are inert or induce only mild elevations in mood and energy when administered to healthy subjects). The insight that raising extracellular dopamine levels is an effect common to both drugs has significantly helped to shape the concept of psychosis as a hyper-dopaminergic state. In this context, it is important to bear in mind that enhanced dopamine release to DAT blockers or releasers is neither specific (see for example [16]) nor a necessary prerequisite for the presence of psychotic symptoms [17, 18]. Thus, the purpose of this and similar research methods is to reshape our understanding of the psychosis-syndrome and to aid stratification of patients for research and clinical purpose according to the underlying pathogenesis.

In contrast to some remarkable but not replicated early findings (see for example [19]), there is now wide agreement on the fact that there are no relevant changes in baseline availability of dopamine D_1 or $D_{2/3}$ receptors in patients with schizophrenia (due to a lack of suitable radioligands to low levels of expression, attempts to image dopamine D_4 and D_5 receptors have not been successful so far). Soon after the introduction of the benzamide $D_{2/3}$ receptor radioligand [^{11}C]raclopride into human PET imaging [20, 21], it was noted that raclopride binding was sensitive towards changes in the concentration of endogenous dopamine in the rodent brain [22, 23]. Initially, this was seen as a possible weakness for reliably quantifying $D_{2/3}$ receptors in the brain. However, the potential of this effect for imaging changes in dopamine levels in the living brain was soon understood [24, 25]. Since then, many studies have shown changes in receptor binding after pharmacological or behavioral manipulation of brain extracellular dopamine levels. In human studies, the most frequently adopted strategies for manipulating extracellular dopamine levels are to pharmacologically induce an increase in extracellular dopamine by administering methylphenidate or d-amphetamine, or to induce a decrease by administering a dopamine-depleting agent such as alpha-methyl para tyrosine (AMPT). Together with evidence from [^{18}F]DOPA PET imaging, PET studies showing enhanced d-amphetamine-induced reductions in $D_{2/3}$ receptor radioligand binding have contributed substantially to the fact that it is now widely accepted that there is enhanced dopamine transmission at least in a large proportion of patients with schizophrenia [1].

3.1 Theoretical Background of "Competition" Studies

Simultaneous measurements of changes in $D_{2/3}$ receptor radioligand binding and extracellular dopamine levels after d-amphetamine have shown a linear relationship between both measures suggesting that increased dopamine is indeed what causes decreased $D_{2/3}$ radioligand binding [26, 27]. These studies suggest that 1% decrease in radioligand binding is indicative of an increase in extracellular dopamine of approximately 40% [28]. However, data in these studies showed large variability, and the exact mechanism leading to these reductions in PET or SPECT experiments is not entirely understood.

While "competition" is frequently used as a cursory explanation, a series of experiments have shown that noncompetitive mechanisms contribute significantly to reductions in $D_{2/3}$ receptor radioligand binding after d-amphetamine administration [29, 30]. Pure competition reduces the affinity ($1/K_D$) of a ligand to a receptor without affecting the maximal number of binding sites (B_{max}). This principle holds true in studies reducing concentration of endogenous dopamine using depleting agents such as reserpine or AMPT [31, 32]. However, some observations on d-amphetamine induced reductions in radioligand binding are difficult to reconcile with a pure competition model. One example is that reductions in $D_{2/3}$ receptor radioligand binding outlast d-amphetamine induced elevations in extracellular dopamine [33, 34].

For disentangling d-amphetamine induced changes in affinity and changes in the number of $D_{2/3}$ receptor binding sites, Ginovart et al. [30] used a Scatchard approach for [^{11}C]raclopride PET in cats. A Scatchard plot is a linearized graphical analysis method depicting the relationship between the concentrations of free and bound ligands in a system. For the in vivo PET approach, Ginovart et al. used [^{11}C]raclopride with high and low specific activity. This study showed changes in $D_{2/3}$ receptor B_{max} *and* K_D after d-amphetamine administration, suggesting that d-amphetamine induced reductions in radioligand binding involve at least two different mechanisms.

In summary, the pharmacology of $D_{2/3}$ receptor "competition" studies is only partially understood, and the method is not suited for measuring absolute levels in extracellular dopamine. However, it yields fairly reliable estimates on relative changes in brain extracellular dopamine after behavioural or pharmacologic interventions.

3.2 Imaging Procedures and Data Analysis in "Competition" Studies

Measuring relative changes in extracellular dopamine implicates the need for a baseline value as comparator for the effects of the intervention. Basically, there are two different approaches used in the literature: One uses two PET scans, one without intervention (baseline) and another one with intervention—for example after d-amphetamine administration. The other approach makes use of a bolus plus constant infusion paradigm, where changes can be imaged in one scanning session.

In a two-scan approach, d-amphetamine is administered orally or intravenously before radioligand injection. These studies are thus pretreatment or blocking paradigms. Extracellular dopamine bound to $D_{2/3}$ receptors blocks radioligand binding and leads to reductions in specific $D_{2/3}$ binding in target ROIs. Attention needs to be paid if d-amphetamine leaves binding in the reference region (usually cerebellum) unaltered. This should be the case, as a prerequisite for a reference region is the absence of specific binding. Kinetic analysis assumes that delivery and washout rates in the reference region are unaltered by d-amphetamine, or at least, that changes in target ROIs match those in the reference region. However, this needs to be ascertained for every radioligand used. In animal experiments, attention needs to be paid on possible interactions between d-amphetamine effects and the effects of anaesthesia. The outcome measure in reference region-based approaches usually is the so-called non-displaceable binding potential (BP_{ND}; [35]) defined as

$$BP_{ND} = B_{max} / K_D$$

where B_{max} indicates the maximal number of available binding sites, while K_D is an inverse measure for the affinity of the radioligand for the receptor. Under tracer conditions, BP_{ND} values are assumed to be linearly proportional to the number of binding sites. However, it is not possible to disentangle changes in affinity and receptor availability in a single PET scan (see above).

Relative changes in radioligand binding are usually reported as percent change in BP_{ND} values and calculated as

$$BP_{ND} = \left(BP_{ND\,d\ amphetamine} - BP_{ND\,baseline}\right) / \left(BP_{ND\,baseline}\right) \times 100$$

As mentioned above, d-amphetamine induced reductions in radioligand binding may last for several hours. It may thus be preferable to perform baseline scans before d-amphetamine scans. On the other hand, order effects, due for example to the effects of novelty and consecutive changes in dopaminergic tone in subjects undergoing PET scanning for their first time are better controlled for in a crossover study design with randomized scan order. In addition, possible carryover effects, i.e., $D_{2/3}$ receptor occupancy by the radioligand itself, should be considered in the study design. A [^{11}C]-(+)-PHNO PET in baboons [36] found significant residual occupancy in dopamine D_3 receptor-rich regions (ventral striatum, pallidum, substantia nigra / ventral tegmental area) at an interscan interval of 3 h. A recent study in humans [37] showed consistently lower [^{11}C]-(+)-PHNO binding in the second scan performed approximately 5 h after the first one. However, reductions were not statistically significant, and no relevant residual occupancy was found in other brain regions.

3.3 Bolus Plus Constant Infusion Paradigms

The outcome measures for quantification of radioligand binding to receptors are based on the assumption of equilibrium binding conditions, or else, a state where the net exchange between the plasma free fraction of the radioligand, nonspecific and specific binding are in steady state. Since this is not the case when the radioligand is injected as a bolus and its concentration changes at differing rates in different tissues, mathematical compartmental modeling is used to infer concentration ratios and volumes of distribution. For some radioligands, however, it is possible to achieve "true" equilibrium conditions by first injecting the ligand as a bolus an then, in the course of the PET scan, supplying the ligand at a rate where concentration ratios between tissues remain constant. This method is usually denoted "bolus plus constant infusion" and can offer a series of advantages over conventional bolus PET imaging, for example simple and reliable quantification of binding parameters.

However, bolus plus constant infusion paradigm is not straightforward and sometimes technically challenging. The exact modalities for a bolus plus constant infusion need to be establishment experimentally for each radioligand, as depending on tracer kinetics, equilibrium binding is not always achieved during the time of a PET scan, especially when using short-lived isotopes such as [^{11}C]. If the concentration ratio is not truly constant but changes at a stable rate during a scan, this is denominated "pseudo-equilibrium" and can lead to errors in the estimates of concentration ratios. For radioligands washing out at differing rates in the different ROIs—as is the case for [^{11}C]-(+)-PHNO (see Fig. 1)—a given amount of radioligand constantly supplied will yield "true" equilibrium in some ROIs and pseudo-equilibrium in other ROIs. Still, even for [^{11}C]-(+)-PHNO, a radioligand known to bind to at least two relevant receptor populations (dopamine D_2 and D_3 receptors) with different kinetics, Lee et al. [38] recently succeeded in developing a bolus plus constant infusion paradigm with good reproducibility and reliable results.

For "competition" studies, the bolus plus constant infusion method has the big advantage that it becomes possible to measure dopamine release in on single scanning session. This reduces radioactivity exposure and time spent in the scanner for research participants and logistical and financial burden for researchers. Moreover, the method has the big advantage that it helps reducing the potentially biasing factors in a two-scan paradigm (various physiological and environmental changes occurring from one day to the other) to a minimum.

4 Notes

4.1 Studying Patients with Psychotic Disorders

Administering d-amphetamine to psychotic patients is sometimes considered to be ethically questionable, as it may temporarily intensify positive symptoms (delusions and hallucinations) and schizophrenic thought disorder. However, psychopathological ratings of symptom severity show an increase of a few percent only, and changes are self-limited in time and return to baseline after a few hours. Nevertheless, studies of this kind require a medical team with profound experience in treating psychotic patients. In our experience, a low dose of d-amphetamine is usually well tolerated also by patients with psychotic disorders. And although positive symptoms and—according to our observations—in particular thought disorder intensify for a few hours, patients usually experience no particular distress under low-dose d-amphetamine. Rather, they report a general improvement in wellbeing. d-Amphetamine doses typically used in these studies are 0.3–0.5 mg/kg bodyweight. d-Amphetamine is administered either orally 1–2 h before scanning (in order to reach maximal d-amphetamine plasma concentrations during the PET scan) or intravenously immediately before, or in case of the so-called bolus plus constant infusion protocols (see above), during the scan.

Changes in positive symptoms are usually measured using the brief psychiatric rating scale (BPRS). The BPRS [39] is a brief scale rating overall psychiatric symptoms and contains specific items for psychotic symptoms. As ratings typically are performed several times within hours, the use of more comprehensive scales such as the positive and negative symptom scale (PANSS) is impractical.

A special challenge is recruitment and selection of patients whose symptoms are characteristic and intense enough for a clear-cut diagnosis (diagnosing schizophrenia can be difficult, especially during early stages of the disorder), but not as severe as to compromise patient safety or full understanding of study procedures and the ability to give informed consent. At therapeutic doses, antipsychotic medication induces significant occupancy of dopamine $D_{2/3}$ receptors (usually 60–80 %; see for example [40]), making it impossible to disentangle occupancy induced by changes in endogenous dopamine and antipsychotic-induced occupancy. Thus, patients need to be either drug-naïve or drug-free for a sufficient period of time (due to slow elimination of certain antipsychotics from $D_{2/3}$ receptor-rich brain regions at least 2 weeks or longer [41]). Moreover, patients should be able to safely tolerate a delay of a few days in antipsychotic treatment, usually imposed by study logistics. Since a significant proportion of patients come to clinical attention several months or more after the first onset of psychotic symptoms, a delay in specific treatment of a few days usually imposes no significant discomfort or clinical risk to patients. Sometimes, however, it is clinically indicated and necessary to administer benzodiazepines for alleviating anxiety or psychomotor agitation. In order to avoid introducing bias, the benzodiazepine dose should be kept at a minimum and stable for the study period.

Acknowledgments

This work was supported by a grant (P23585) of the FWF Austrian Science Fund granted to M.W.

References

1. Howes OD, Kambeitz J, Kim E, Stahl D, Slifstein M, Abi-Dargham A, Kapur S (2012) The nature of dopamine dysfunction in schizophrenia and what this means for treatment. Arch Gen Psychiatry 69(8):776–786. doi:10.1001/archgenpsychiatry.2012.169
2. Bauer M, Praschak-Rieder N, Kasper S, Willeit M (2012) Is dopamine neurotransmission altered in prodromal schizophrenia? A review of the evidence. Curr Pharm Des 18(12):1568–1579
3. Wilson AA, McCormick P, Kapur S, Willeit M, Garcia A, Hussey D, Houle S, Seeman P, Ginovart N (2005) Radiosynthesis and evaluation of [11C]-(+)-4-propyl-3,4,4a,5,6,10b-hexahydro-2H-naphtho[1,2-b][1,4]oxazin-9-ol as a potential radiotracer for in vivo imaging of the dopamine D2 high-affinity state with positron emission tomography. J Med Chem 48(12):4153–4160. doi:10.1021/jm050155n
4. Lammertsma AA, Hume SP (1996) Simplified reference tissue model for PET receptor studies. Neuroimage 4(3 Pt 1):153–158. doi:10.1006/nimg.1996.0066
5. Ginovart N, Willeit M, Rusjan P, Graff A, Bloomfield PM, Houle S, Kapur S, Wilson AA (2007) Positron emission tomography quantification of [11C]-(+)-PHNO binding in the human brain. J Cereb Blood Flow Metab 27(4):857–871. doi:10.1038/sj.jcbfm.9600411
6. Cumming P (2009) Imaging dopamine. Cambridge University Press, Cambridge
7. Gunn RN, Lammertsma AA, Hume SP, Cunningham VJ (1997) Parametric imaging of ligand-receptor binding in PET using a simplified reference region model. Neuroimage 6(4):279–287. doi:10.1006/nimg.1997.0303
8. Oldendorf WH, Szabo J (1976) Amino acid assignment to one of three blood-brain barrier amino acid carriers. Am J Physiol 230(1):94–98
9. Kumakura Y, Cumming P, Vernaleken I, Buchholz HG, Siessmeier T, Heinz A, Kienast T, Bartenstein P, Grunder G (2007) Elevated [18F]fluorodopamine turnover in brain of patients with schizophrenia: an [18F]fluorodopa/positron emission tomography study. J Neurosci 27(30):8080–8087. doi:10.1523/JNEUROSCI.0805-07.2007
10. Schabram I, Henkel K, Mohammadkhani Shali S, Dietrich C, Schmaljohann J, Winz O, Prinz S, Rademacher L, Neumaier B, Felzen M, Kumakura Y, Cumming P, Mottaghy FM, Grunder G, Vernaleken I (2014) Acute and sustained effects of methylphenidate on cognition and presynaptic dopamine metabolism: an [18F]FDOPA PET study. J Neurosci 34(44):14769–14776. doi:10.1523/JNEUROSCI.1560-14.2014
11. Hartvig P, Tedroff J, Lindner KJ, Bjurling P, Chang CW, Tsukada H, Watanabe Y, Langstrom B (1993) Positron emission tomographic studies on aromatic L-amino acid decarboxylase activity in vivo for L-dopa and 5-hydroxy-L-tryptophan in the monkey brain. J Neural Transm Gen Sect 94(2):127–135
12. Bruck A, Aalto S, Nurmi E, Vahlberg T, Bergman J, Rinne JO (2006) Striatal subregional 6-[18F]fluoro-L-dopa uptake in early Parkinson's disease: a two-year follow-up study. Mov Disord 21(7):958–963. doi:10.1002/mds.20855
13. Hoshi H, Kuwabara H, Leger G, Cumming P, Guttman M, Gjedde A (1993) 6-[18F]fluoro-L-dopa metabolism in living human brain: a comparison of six analytical methods. J Cereb Blood Flow Metab 13(1):57–69. doi:10.1038/jcbfm.1993.8
14. Ishiwata K, Kawamura K, Yanai K, Hendrikse NH (2007) In vivo evaluation of P-glycoprotein modulation of 8 PET radioligands used clinically. J Nucl Med 48(1):81–87
15. Lieberman JA, Kane JM, Alvir J (1987) Provocative tests with psychostimulant drugs in schizophrenia. Psychopharmacology (Berl) 91(4):415–433
16. Boileau I, Payer D, Chugani B, Lobo DS, Houle S, Wilson AA, Warsh J, Kish SJ, Zack M (2014) In vivo evidence for greater amphetamine-induced dopamine release in pathological gambling: a positron emission tomography study with [(11)C]-(+)-PHNO. Mol Psychiatry 19(12):1305–1313. doi:10.1038/mp.2013.163
17. Kapur S (2003) Psychosis as a state of aberrant salience: a framework linking biology, phenomenology, and pharmacology in schizophrenia. Am J Psychiatry 160(1):13–23
18. Howes OD, Kapur S (2009) The dopamine hypothesis of schizophrenia: version III—the final common pathway. Schizophr Bull 35(3):549–562. doi:10.1093/schbul/sbp006, sbp006 [pii]
19. Wong DF, Wagner HN Jr, Tune LE, Dannals RF, Pearlson GD, Links JM, Tamminga CA, Broussolle EP, Ravert HT, Wilson AA, Toung JK, Malat J, Williams JA, O'Tuama LA, Snyder SH, Kuhar MJ, Gjedde A (1986) Positron

emission tomography reveals elevated D2 dopamine receptors in drug-naive schizophrenics. Science 234(4783):1558–1563
20. Kohler C, Hall H, Ogren SO, Gawell L (1985) Specific in vitro and in vivo binding of 3H-raclopride. A potent substituted benzamide drug with high affinity for dopamine D-2 receptors in the rat brain. Biochem Pharmacol 34(13):2251–2259
21. Farde L, Hall H, Ehrin E, Sedvall G (1986) Quantitative analysis of D2 dopamine receptor binding in the living human brain by PET. Science 231(4735):258–261
22. Seeman P, Guan HC, Niznik HB (1989) Endogenous dopamine lowers the dopamine D2 receptor density as measured by [3H]raclopride: implications for positron emission tomography of the human brain. Synapse 3(1):96–97. doi:10.1002/syn.890030113
23. Ross SB, Jackson DM (1989) Kinetic properties of the accumulation of 3H-raclopride in the mouse brain in vivo. Naunyn Schmiedebergs Arch Pharmacol 340(1):6–12
24. Farde L, Nordstrom AL, Wiesel FA, Pauli S, Halldin C, Sedvall G (1992) Positron emission tomographic analysis of central D1 and D2 dopamine receptor occupancy in patients treated with classical neuroleptics and clozapine. Relation to extrapyramidal side effects. Arch Gen Psychiatry 49(7):538–544
25. Innis RB, Malison RT, al-Tikriti M, Hoffer PB, Sybirska EH, Seibyl JP, Zoghbi SS, Baldwin RM, Laruelle M, Smith EO et al (1992) Amphetamine-stimulated dopamine release competes in vivo for [123I]IBZM binding to the D2 receptor in nonhuman primates. Synapse 10(3):177–184. doi:10.1002/syn.890100302
26. Breier A, Su TP, Saunders R, Carson RE, Kolachana BS, de Bartolomeis A, Weinberger DR, Weisenfeld N, Malhotra AK, Eckelman WC, Pickar D (1997) Schizophrenia is associated with elevated amphetamine-induced synaptic dopamine concentrations: evidence from a novel positron emission tomography method. Proc Natl Acad Sci U S A 94(6):2569–2574
27. Laruelle M, Iyer RN, al-Tikriti MS, Zea-Ponce Y, Malison R, Zoghbi SS, Baldwin RM, Kung HF, Charney DS, Hoffer PB, Innis RB, Bradberry CW (1997) Microdialysis and SPECT measurements of amphetamine-induced dopamine release in nonhuman primates. Synapse 25(1):1–14. doi:10.1002/(SICI)1098-2396(199701)25:1<1::AID-SYN1>3.0.CO;2-H
28. Laruelle M (2000) Imaging synaptic neurotransmission with in vivo binding competition techniques: a critical review. J Cereb Blood Flow Metab 20(3):423–451
29. Sun W, Ginovart N, Ko F, Seeman P, Kapur S (2003) In vivo evidence for dopamine-mediated internalization of D2-receptors after amphetamine: differential findings with [3H] raclopride versus [3H]spiperone. Mol Pharmacol 63(2):456–462
30. Ginovart N, Wilson AA, Houle S, Kapur S (2004) Amphetamine pretreatment induces a change in both D2-Receptor density and apparent affinity: a [11C]raclopride positron emission tomography study in cats. Biol Psychiatry 55(12):1188–1194. doi:10.1016/j.biopsych.2004.02.019
31. Ginovart N, Farde L, Halldin C, Swahn CG (1997) Effect of reserpine-induced depletion of synaptic dopamine on [11C]raclopride binding to D2-dopamine receptors in the monkey brain. Synapse 25(4):321–325. doi:10.1002/(SICI)1098-2396(199704)25:4<321::AID-SYN2>3.0.CO;2-C
32. Laruelle M, D'Souza CD, Baldwin RM, Abi-Dargham A, Kanes SJ, Fingado CL, Seibyl JP, Zoghbi SS, Bowers MB, Jatlow P, Charney DS, Innis RB (1997) Imaging D2 receptor occupancy by endogenous dopamine in humans. Neuropsychopharmacology 17(3):162–174. doi:10.1016/S0893-133X(97)00043-2
33. Narendran R, Slifstein M, Hwang DR, Hwang Y, Scher E, Reeder S, Martinez D, Laruelle M (2007) Amphetamine-induced dopamine release: duration of action as assessed with the D2/3 receptoragonist radiotracer (-)-N-[(11) C]propyl-norapomorphine ([11C]NPA) in an anesthetized nonhuman primate. Synapse 61(2):106–109. doi:10.1002/syn.20346
34. Cardenas L, Houle S, Kapur S, Busto UE (2004) Oral D-amphetamine causes prolonged displacement of [11C]raclopride as measured by PET. Synapse 51(1):27–31. doi:10.1002/syn.10282
35. Innis RB, Cunningham VJ, Delforge J, Fujita M, Gjedde A, Gunn RN, Holden J, Houle S, Huang SC, Ichise M, Iida H, Ito H, Kimura Y, Koeppe RA, Knudsen GM, Knuuti J, Lammertsma AA, Laruelle M, Logan J, Maguire RP, Mintun MA, Morris ED, Parsey R, Price JC, Slifstein M, Sossi V, Suhara T, Votaw JR, Wong DF, Carson RE (2007) Consensus nomenclature for in vivo imaging of reversibly binding radioligands. J Cereb Blood Flow Metab 27(9):1533–1539. doi:10.1038/sj.jcbfm.9600493
36. Girgis RR, Xu X, Miyake N, Easwaramoorthy B, Gunn RN, Rabiner EA, Abi-Dargham A, Slifstein M (2011) In vivo binding of antipsy-

chotics to D3 and D2 receptors: a PET study in baboons with [11C]-(+)-PHNO. Neuropsychopharmacology 36(4):887–895. doi:10.1038/npp.2010.228

37. Gallezot JD, Zheng MQ, Lim K, Lin SF, Labaree D, Matuskey D, Huang Y, Ding YS, Carson RE, Malison RT (2014) Parametric imaging and test-retest variability of (1)(1) C-(+)-PHNO binding to D(2)/D(3) dopamine receptors in humans on the high-resolution research tomograph PET scanner. J Nucl Med 55(6):960–966. doi:10.2967/jnumed.113.132928
38. Lee DE, Gallezot JD, Zheng MQ, Lim K, Ding YS, Huang Y, Carson RE, Morris ED, Cosgrove KP (2013) Test-retest reproducibility of [11C]-(+)-propyl-hexahydro-naphtho-oxazin positron emission tomography using the bolus plus constant infusion paradigm. Mol Imaging 12(2):77–82
39. Overall JE, Gorham DR (1962) The brief psychiatric rating scale. Psychol Rep 10:799–812
40. Bishara D, Olofinjana O, Sparshatt A, Kapur S, Taylor D, Patel MX (2013) Olanzapine: a systematic review and meta-regression of the relationships between dose, plasma concentration, receptor occupancy, and response. J Clin Psychopharmacol 33(3):329–335. doi:10.1097/JCP.0b013e31828b28d5
41. Tauscher J, Jones C, Remington G, Zipursky RB, Kapur S (2002) Significant dissociation of brain and plasma kinetics with antipsychotics. Mol Psychiatry 7(3):317–321. doi:10.1038/sj.mp.4001009

Index

Heinz Bönisch and Harald H. Sitte (eds.), *Neurotransmitter Transporters: Investigative Methods*, Neuromethods, vol. 118, DOI 10.1007/978-1-4939-3765-3, © Springer Science+Business Media New York 2016

C

D

I

J

K

L

M

N

O

W

X

Z

Printed in the United States
By Bookmasters